STUDY GUIDE

STUDY GUIDE

Judith Stewart
Community College of Southern Nevada

Human Biology

Concepts and Current Issues

SIXTH EDITION

Michael D. Johnson

Benjamin Cummings

Boston   Columbus   Indianapolis   New York   San Francisco   Upper Saddle River
Amsterdam   Cape Town   Dubai   London   Madrid   Milan   Munich   Paris   Montréal   Toronto
Delhi   Mexico City   São Paulo   Sydney   Hong Kong   Seoul   Singapore   Taipei   Tokyo

Editor-in-Chief: Beth Wilbur
Executive Director of Development: Deborah Gale
Acquisitions Editor: Becky Ruden
Associate Editor: Brady Golden
Assistant Editor: Leslie Allen
Executive Marketing Manager: Lauren Harp
Executive Managing Editor: Erin Gregg
Managing Editor: Michael Early
Senior Production Project Manager: Shannon Tozier
Production Service and Composition: PreMedia Global
Illustrations: Imagineering
Cover Production: Seventeenth Street Studio
Manufacturing Buyer: Michael Penne

Cover Image: Zap Art/Getty Images

ISBN 10: 0-321-74251-6
ISBN 13: 978-0-321-74251-3

8 9 10 CP 14 13 12

Printed in Canada

Benjamin Cummings
is an imprint of

www.pearsonhighered.com

Contents

Top 10 Study Tips for Human Biology

As you begin your study of science, one of the most important skills to acquire is the belief that you can succeed. Commitment, adequate study time, practice, improving your study skills, and utilizing available resources are the foundation for mastering the concepts in your Textbook. Combined with the study skills described below, using the *Study Guide* can increase the confidence with which you approach your assignments and exams.

To evaluate your study skills, consider:

1. **Environment**—Have you set aside a quiet place and time to study every day? This may be a table in the campus library or your kitchen table during a quiet time of the day. Keep distractions to a minimum and turn off the television and any music with lyrics that draw your thoughts away from what you're studying.

2. **Consistency**—You get the greatest return on your investment of study time when you study every day. Short daily study sessions are more effective than long study sessions that occur only once a week.

3. **Commitment**—There is no shortcut to learning biology. You should plan to spend at least two to three hours studying outside of class for every hour you spend in class. Schedule blocks of study time into your calendar every week.

4. **Focus**—Distracting thoughts often pop into your mind when studying, stealing your valuable time. Keep a small pad of paper nearby while you study and jot down these thoughts; this gives your mind permission to let them go, knowing that you can revisit them later when your study time is over.

5. **Prioritize**—With many demands on your time, you must set aside time to study. This may be easier if you break it into smaller tasks. For example, you might decide to arrive for class 10 minutes early and use those 10 minutes to review your Textbook or practice with the *Study Guide*. Or you might assign one hour every Saturday morning to take and review the Chapter Test.

6. **Adaptations**—Adapt the exercises in the *Study Guide* to work with your schedule and your individual needs. For example, if you have short study opportunities "on the go", convert the Matching, Fill-In-The-Blank, and Short Answer exercises to flash cards bound by a ring clip.

7. **Prepare**—Allocate some of your study time each week to preparing for class lectures. Read the assigned material ahead of time and come to class with questions. Listen for the answers during the lecture, or ask your instructor before you leave class.

8. **Become an Active Reader**—It is easy to spend an hour highlighting a chapter in your textbook and feel productive, but if you don't remember what you read the hour has been wasted. Limit your highlighting to 2 or 3 key words at a time. Stop after each section in your reading to be sure you can repeat the key concepts you just read; if you cannot, go back and reread that section. Read with a questioning mind. You should have questions as you read; write them on a pad of paper or in the margins of your textbook, then watch for answers as you continue your reading.

9. **Assess, Assess, Assess**—Memorization is a beginning step in human biology. However, it cannot provide the level of understanding that you need to succeed. Practice is critical, and self-assessment lets you find your weak areas before an exam. Use the Chapter Tests and Self Tests as "pre-tests." Do these exercises as if you were taking an exam, without your notes or book. Then use the keys to "grade" the pretest, and spend additional study time on your weakest areas.

10. **Be Accountable**—Take responsibility for your own learning. You are the only person who can do this for yourself and only you have power over your own actions.

Notes on Using Your *Study Guide*

Each chapter of the *Study Guide* includes the following components:

Chapter Summary and Objectives: A list of 20 Key Concepts, from the textbook.

Exercises: A combination of the following types of questions test your understanding of the topic: color-coding, figure-labeling, essay, short answer, crossword puzzles, definitions, matching, fill-in-the-blank, completion, word choice, and true/false. Answers to most exercises are found in the Answer Key, at the back of the *Study Guide*.

Looking Back/Looking Forward: Questions include information from previous and future chapters, providing an opportunity for you to incorporate material already learned and anticipate future material.

Chapter Test: A selection of 25 multiple-choice questions assess your mastery of the topic and predict your performance on an exam. Take the Chapter Test only *after* you have successfully completed the exercises.

Self Test: The 20 Key Concepts found at the beginning of the *Study Guide* chapter are rewritten as Self-Test questions to assess your comprehension of the main topics in the chapter. You can find answers to these questions by referring back to the Chapter Summary and Objectives for that chapter.

Learning science can change your life, your perspective on many societal issues, and your ability to make informed choices. Establish a study routine that works for you, and you're on your way to a successful course.

JUDY STEWART

1

Human Biology, Science, and Society

Chapter Summary and Key Concepts

After reading and studying this chapter you should know the following:

Sections 1.1, 1.2

1. Biology is the study of life.

2. Living things can be characterized by their molecular composition, energy requirements, cellular nature, homeostatic mechanisms, responsiveness, ability to grow and reproduce, and evolution.

3. A system of classification categorizes living things according to their characteristics.

4. As living organisms, humans are characterized by bipedalism, opposable thumbs, a large brain, and complex language skills.

5. Levels of biological organization exist in living things, beginning with atoms and becoming progressively more complex and inclusive.

Sections 1.3, 1.4

6. Scientific knowledge is information about the natural world and is gained by practicing the scientific method.

7. The scientific method is a series of steps that allow scientists to gather information about the natural world and test their ideas.

8. The scientific method begins with an observation and is followed by a hypothesis, a prediction, an experiment, data, and a conclusion.

9. A well-supported hypothesis becomes a theory.

10. Scientists publish their work in peer-reviewed journals to allow other scientists to verify their findings.

11. Scientific information is presented in many forms; it is important to be aware of different degrees of accuracy in each type of media.

Sections 1.5, 1.6

12. Good scientists use the tools of critical thinking.

13. Informed individuals can become critical thinkers.

14. Science uses statistics to interpret data.

15. A graph is one effective way to present information.

16. Facts differ from anecdotes and conclusions because facts are verifiable.

17. Correlation is a pattern between two variables, and it does not necessarily indicate causality.

18. Technology is the application of science and has great potential to improve our world.

19. Science is an explanation of observable events in a physical world; it cannot be used to explain or support feelings, opinions, or abstract ideas.

20. Knowledge of science helps individuals to make informed choices.

Exercises

Complete the exercises for each section after you have read and studied the section. If you cannot answer some questions, or if you answer them incorrectly, return to the chapter and review this information. You may find it helpful to work on only one section at a time. When you have completed all sections, take the Chapter Test as an indicator of your mastery of this topic.

1.1   The characteristics of life

1.2   How humans fit into the natural world

Define

Write the term(s) for each definition:

_____   1.   the study of the natural world

_____   2.   the study of living organisms and life's processes

_____   3.   physical and chemical processes that transform energy to maintain life

_____   4.   the maintenance of a relatively constant internal environment

_____   5.   this type of organism is composed of many cells or many different types of cells

_____   6.   the genetic material of cells

_____   7.   in the five-kingdom classification system, this kingdom contains all prokaryotic organisms

_____ 8. this kingdom contains multicellular animals

_____ 9. this kingdom includes multicellular plants

_____ 10. this kingdom contains multicellular decomposers

_____ 11. this kingdom includes unicellular eukaryotic organisms

_____ 12. in the three-domain classification system, this domain includes all eukaryotic organisms

_____ 13. the two domains that contain all prokaryotic organisms

_____ 14. one or more populations of organisms with similar characteristics and capable of breeding and producing fertile offspring under natural conditions

_____ 15. the second smallest unit of classification

Fill-in-the-Blank

Referenced sections are indicated in parentheses.

16. Only living things are able to create the complex molecules of life: _____, _____, _____, and _____. (1.1)

17. Plants convert the energy of sunlight into energy they can use in the process of _____. (1.1)

18. The natural world includes all _____ and all _____. (1.1)

19. _____ cells have a nucleus, and _____ cells lack a nucleus. (1.1)

20. When exhibiting the seven signs of life, both living and nonliving things can grow, but only living things can _____. (1.1)

21. Animals with a nerve cord and a backbone are called _____. (1.2)

22. Vertebrates with mammary glands are classified as _____. (1.2)

23. Humans, apes, and monkeys are all considered mammals and are further classified as _____. (1.2)

24. The ability to stand upright and walk on two legs is called _____. (1.2)

25. Humans have _____ thumbs, allowing them to grasp small objects with precision. (1.2)

26. Humans have a large _____ mass in relation to their body size. (1.2)

27. The capacity for complex _____ is a defining feature of being human. (1.2)

Ordering

Place the following levels of biological organization in the proper sequence, beginning with the simplest and ending with the most complex.

tissue   organ system   population

organism   community   atom

cell   organ   ecosystem

biosphere   molecule

28. _____ 29. _____ 30. _____

31. _____ 32. _____ 33. _____

34. _____ 35. _____ 36. _____

37. _____ 38. _____

Short Answer

39. List the seven criteria accepted by scientists as characteristics of life.

40. When are the characteristics of life not applicable in describing living things?

41. How is a domain different from a kingdom?

Critical Thinking

42. An organelle is a small functional unit found within a cell. Why is a single organelle not considered to be alive?

43. In order to sustain life, human beings obtain energy and raw materials from both living and nonliving things. Provide an example of how you, as a human being, exhibit this characteristic each day.

44. Maintaining homeostasis and responding to the external environment are two characteristics of living things. How are these two characteristics similar and how are they different?

1.3 Science is both a body of knowledge and a process

1.4 Sources of scientific information vary in style and quality

1.5 Learning to be a critical thinker

1.6 The role of science in society

Matching

Select from the following list the correct term for each definition below.

scientific knowledge	**hypothesis**	**placebo**
scientific method	**experiment**	**variable**
experimental group	**skepticism**	**theory**
control group	**technology**	**data**

_____ 1. a broad hypothesis that has been repeatedly tested and supported over time

_____ 2. a false treatment given to a control group in an experiment

_____ 3. a factor that might vary during the course of an experiment

_____ 4. the application of science

_____ 5. the group in a controlled experiment that is not exposed to the experimental treatment

_____ 6. a questioning attitude required to practice science

_____ 7. organized, reliable information about the natural world

_____ 8. a tentative statement about the natural world

_____ 9. a carefully planned and executed test of a prediction

_____ 10. the group in a controlled experiment that is exposed to the experimental treatment

_____ 11. the process used to acquire scientific knowledge

_____ 12. numerical information obtained from an experiment

True/False

Mark each statement true or false. If a statement is marked false, replace the underlined term or phrase with the term or phrase required to make the statement true. Referenced sections are indicated in parentheses.

_____ 13. Science is a study of the natural world and includes both knowledge and <u>speculation</u>. (1.3)

_____ 14. When we use observations to make broad generalizations, we are using <u>deductive</u> reasoning. (1.3)

_____ 15. A hypothesis is a statement that can lead to <u>testable</u> deductions. (1.3)

_____ 16. A working hypothesis is a testable <u>theory</u>. (1.3)

_____ 17. The variable of interest in an experiment is called the <u>controlled</u> variable. (1.3)

_____ 18. When the subjects in an experiment do not know who is in the control group and who is in the experimental group, this is a <u>double-blind</u> study. (1.3)

_____ 19. When a prediction is found to be false, the <u>results</u> must be modified. (1.3)

_____ 20. A hypothesis cannot be proved, it can only be <u>disproved</u>. (1.3)

_____ 21. Scientists publish their findings in <u>public</u> journals so that other scientists can examine their work. (1.3)

_____ 22. The horizontal axis of a graph is called the <u>ordinate</u>. (1.5)

_____ 23. The dependent variable is usually plotted on the <u>vertical</u> axis.

_____ 24. The range above and below a numerical average that scientists believe represents accurate data is called the <u>standard error</u>. (1.5)

_____ 25. Information that can be verified is a <u>conclusion</u>. (1.5)

_____ 26. A correlation between two variables does not always mean that one variable <u>caused</u> the other. (1.5)

_____ 27. Science is limited to the explanation of <u>proven</u> events in the natural world. (1.6)

Completion

Scientific information is reported through many types of media. For each type listed below, name one strength and one weakness of this resource for a general biology student. (There are no answers to this exercise in the Answer Key. Refer to Section 1.4 of your text to check your responses.)

Type of Media	Strengths	Weaknesses
Scientific journal		
Popular science magazine		
Newspaper		
Web site ending in .com		
Web site ending in .edu		

Looking Back/Looking Forward

1. Chapter 4 discusses a control mechanism called "negative feedback" in which homeostasis is maintained when sensors detect a change, and responses are initiated to bring the system back to normal. This occurs in temperature regulation. If you are in cold weather without a coat, sensors in skin and internal organs detect a temperature change and send this information to your brain. The brain responds by affecting blood vessels and muscles in such a way that your body temperature increases toward normal levels. Which characteristics of life are demonstrated in this response?

Chapter Test

Multiple Choice

1. Biology is the study of
   a. science.
   b. ecosystems.
   c. life.
   d. biochemistry.

2. Which of the following is the most accurate and inclusive list of what is contained in the natural world?
   a. all matter and all living things
   b. all matter and all nonliving things
   c. all energy and all living things
   d. all matter, all energy, all living things, and all nonliving things

3. Living things require energy. The processes involved in acquiring and transforming energy are collectively referred to as
   a. chemistry.
   b. physical biochemistry.
   c. metabolism.
   d. catabolism.

4. The absence of _____ would indicate a substance is not a living organism.
   a. multicellular structure
   b. molecules of life
   c. a nucleus
   d. external movement

5. What is the smallest level of order at which evolution occurs?
   a. organism
   b. cell
   c. population
   d. ecosystem

6. Classification of organisms in the five kingdom systems depends on all of the following *except*

   a.   the presence or absence of a nucleus.

   b.   the type of metabolism.

   c.   the number of cells.

   d.   the type of reproduction.

7. Which of the following is a domain that is described incorrectly?

   a.   *Prokarya*—contains all prokaryotic organisms

   b.   *Eukarya*—contains all eukaryotic organisms

   c.   *Archaea*—contains primitive unicellular eukaryotes

   d.   *Bacteria*—contains unicellular prokaryotes

8. The defining features of humans include all of the following *except*

   a.   opposable thumbs.

   b.   bipedalism.

   c.   sexual reproduction.

   d.   large brain mass in relation to body size.

9. A group of tissues that carries out a specific function constitutes a(n)

   a.   cell.

   b.   organism.

   c.   organ.

   d.   community.

10. Place the following steps of the scientific method in the correct sequence in which they would be used.

    I.   formulate a hypothesis

    II.   observe and generalize

    III. experiment

    IV.   make a testable prediction

    V.   modify hypothesis if necessary

    a.   I, V, II, IV, III

    b.   II, I, IV, III, V

    c.   II, IV, III, I, V

    d.   IV, I, II, III, V

11. In a scientific experiment, a placebo is given to the

    a.   experimental group.

    b.   control group.

    c.   both groups at different times.

    d.   neither group.

12. You wish to test the effect of different amounts of a particular fertilizer on the growth of tomato plants. In your experiment, the controlled variable is

    a.   the type of fertilizer.

    b.   plant growth.

    c.   the amount of fertilizer.

    d.   the type of tomato plant used.

13. The most accurate results will be obtained by a scientific study that
    a. has a control group and an experimental group.
    b. is double-blind.
    c. is blind.
    d. identifies variables that may affect the outcome of the experiment.

14. Which of the following does not describe a prediction?
    a. an "if . . . then" statement
    b. a vague, general statement
    c. a testable statement
    d. all of the above describe a prediction

15. Confidence in a hypothesis is developed when
    a. many scientists propose the hypothesis at the same time.
    b. the experiment designed to test the hypothesis has a control group and an experimental group.
    c. the results of an experiment support the hypothesis.
    d. many scientists have tried and failed to disprove the hypothesis.

16. The goal of science is to _____ what is _____.
    a. prove, true
    b. prove, false
    c. disprove, true
    d. disprove, false

17. Scientists most often publish their findings
    a. in popular magazines.
    b. in peer-reviewed journals.
    c. in documentaries.
    d. on the Internet.

18. A conclusion is
    a. a fact.
    b. an interpretation.
    c. verifiable information.
    d. a judgment based on fact.

19. Statistics is the mathematics of organizing and interpreting
    a. graphs.
    b. experiments.
    c. variables.
    d. data.

20. You begin eating broccoli every night of the week, and notice that you are more energetic. Which of the following would be the most accurate statement?
    a. Eating broccoli causes you to be more energetic.
    b. Eating broccoli is a good way to combat fatigue.
    c. Eating broccoli appears to be correlated with increased energy.
    d. Eating broccoli provides nutrients that are causing you to have more energy.

21. Developing the skills of a critical thinker will most help you to
    a. understand information.
    b. question information.
    c. generate information.
    d. prove information.

22. Science cannot be used to explain
    a. data.
    b. observations of the physical world.
    c. nonphysical events outside the natural world.
    d. science can be used to explain all of the above.

23. Anecdotal evidence often appears as
    a. a double-blind study.
    b. a testimonial by a nonscientist.
    c. data.
    d. statistical evidence.

24. In a graph, the standard error is represented by
    a. the ordinate.
    b. the abscissa.
    c. small lines that extend above and below the average number.
    d. bars or dots indicating the average of all data collected.

25. Understanding science and the scientific method can help individuals
    a. make informed decisions.
    b. contribute to society.
    c. understand the impact of technology.
    d. evaluate the validity of nonphysical events that do not fall within the limits of the natural world.

Self Test

Each of the Key Concepts listed at the beginning of this chapter has been rewritten as a question below. After successfully completing the study guide exercises and the Chapter Test, you should be able to answer each of these questions. Refer to the Key Concepts list at the beginning of this chapter to check your answers.

1. What is biology?

2. What are the seven characteristics of living things?

3. How do classification systems categorize living things?

4. What are the characteristics of humans?

5. What are the levels of biological organization?

6. What is scientific knowledge, and how is it gained?

7. What is the scientific method?

8. What are the steps of the scientific method?

9. What happens to a well-supported hypothesis?

10. Where do scientists most often publish their work?

11. When reading or watching the presentation of scientific information, what should you be aware of?

12. What type of thinking is used by good scientists?

13. Is the type of thinking you chose in question 12 limited to scientists?

14. What is the value of statistics to science?

15. What is a graph?

16. How do facts differ from anecdotes and conclusions?

17. What is a correlation? How does it relate to causality?

18. What is technology? What is the value of technology?

19. What are the limitations of science?

20. How is knowledge of science valuable to individuals?

2

The Chemistry of Living Things

Chapter Summary and Key Concepts

After reading and studying this chapter you should know the following:

Sections 2.1, 2.2

1. Matter is composed of elements; elements are composed of atoms; atoms are composed of protons, neutrons, and electrons.

2. The atomic number represents the number of protons in an atom; the mass number represents the total number of protons and neutrons in an atom's nucleus.

3. Isotopes are atoms of an element that have more or fewer neutrons than typical atoms.

4. Energy is the ability to do work. Electrons have potential energy based on their position in an electron shell.

5. Chemical bonds form between atoms that need to fill their outermost occupied electron shell.

6. Covalent bonds form when atoms share electrons; ionic bonds form between oppositely charged ions; hydrogen bonds form between polar molecules.

Sections 2.3, 2.4

7. Water is an effective biological solvent because the water molecules are polar, and water is a liquid at body temperature.

8. Water helps to regulate body temperature because it can absorb and hold heat.

9. An acid is a molecule that donates H^+ to a solution. A base is a molecule that absorbs H^+ in a solution.

10. The pH scale is a measure of the hydrogen ion concentration of a solution.

11. A buffer helps to minimize changes in pH.

Sections 2.5, 2.6, 2.7, 2.8, 2.9, 2.10

12. Organic molecules are molecules that contain carbon covalently bonded to other elements.

13. Dehydration synthesis creates macromolecules by removing water. Hydrolysis breaks down macromolecules by adding water.

14. The four classes of organic molecules synthesized by living organisms are carbohydrates, lipids, proteins, and nucleic acids.

15. Carbohydrates in the form of simple sugars provide immediate energy to the cells, while polysaccharides provide stored energy.

16. Lipids include triglycerides that store energy, phospholipids that function in the cell membrane, and steroids that are classified as lipids because they are not soluble in water.

17. Proteins are composed of amino acids and are described at four structural levels.

18. Enzymes are proteins that function as catalysts to speed up chemical reactions.

19. Nucleic acids are composed of nucleotides.

20. DNA and RNA are nucleic acids that function in storing and expressing genetic information. ATP is a nucleic acid that functions as an energy molecule in the cell.

Exercises

Complete the exercises for each section after you have read and studied the section. If you cannot answer some questions, or if you answer them incorrectly, return to the chapter and review this information. You may find it helpful to work on only one section at a time. When you have completed all sections, take the Chapter Test as an indicator of your mastery of this topic.

2.1   All matter consists of elements

2.2   Atoms combine to form molecules

Define

Write the term(s) for each definition below:

_____ 1. anything that has mass and occupies space

_____ 2. a fundamental form of matter that cannot be broken down to a simpler form

_____ 3. the smallest unit of any element that retains the physical and chemical properties of that element.

_____ 4. a positive particle found within the nucleus of an atom

_____ 5. a neutral particle found within the nucleus of an atom

_____ 6. a negatively charged particle found orbiting the nucleus of an atom

_____ 7. similar to weight but not dependent on gravity

_____ 8. a number characteristic of an atom and based on the number of protons in the nucleus

_____ 9. the total number of protons and neutrons in the nucleus of an atom

_____ 10. atoms of the same element with differing numbers of neutrons

_____ 11. isotopes that give off energy in the form of radiation

_____ 12. a stable association between two or more atoms

_____ 13. the capacity to do work

_____ 14. stored energy that is not performing work at the moment

_____ 15. energy that is doing work

_____ 16. attractive forces that bind atoms to each other

_____ 17. a charged atom formed when an atom gains or loses electrons

Fill-in-the-Blank

Referenced sections are indicated in parentheses.

18. There are just over _____ known elements. (2.1)

19. All known elements are arranged into groups according to their similar properties, in a table known as the _____ table.

20. The smallest unit of matter that can take part in a chemical reaction is a(n) _____.

21. The second shell of an atom can hold up to _____ electrons. (2.1)

22. All atoms of the same element have the same number of _____. (2.1)

23. The chemical formula H_2S indicates that one molecule of this substance will have _____ atom(s) of hydrogen and _____ atom(s) of sulfur. (2.1)

24. If an electron moves from the 3rd shell of an atom to the 2nd shell, will it gain or lose energy? _____. (2.2)

25. When atoms share electrons in order to fill their outermost occupied electron shells, a(n) _____ bond is formed. (2.2)

26. When a chemical bond forms between oppositely charged ions, it will be a(n) _____ bond. (2.2)

27. Ions in aqueous solution are called electrolytes because these solutions are able to _____ _____. (2.2)

Completion

Fill in the missing information for each atom or ion in the table.

Element	Atomic Number	Atomic Mass	Number of Protons	Number of Neutrons	Number of Electrons
28. Na	11	23	a)	b)	c)
29. Na⁺	a)	23	11	b)	c)
30. P	a)	b)	15	16	c)
31. Mg	a)	b)	c)	12	12
32. H	1	1	a)	b)	c)

Critical Thinking

33. Do double bonds ever occur between atoms joined by ionic bonds? Explain your answer.

34. a. What types of chemical bonds will exist in a glass of pure water?

    b. What types of chemical bonds will exist in a glass of salt water if some of the salt remains undissolved?

35. Is H_2 a polar molecule? Explain your answer.

2.3   Life depends on water
2.4   The importance of hydrogen ions

Matching

Select from the following list the correct term for each definition below.

solvent   hydrophilic   buffer   acid

base   hydrophobic   pH scale   solute

_____ 1. a molecule that can give up a hydrogen ion

_____ 2. a molecule that can accept a hydrogen ion

_____ 3. any substance that minimizes changes in the pH of a solution when an acid or base is added

_____ 4. molecules that are polar and are attracted to water

_____ 5. a dissolved substance

_____ 6. molecules that are not polar and are not attracted to water

_____ 7. a liquid in which other substances dissolve

_____ 8. a measure of H^+ concentration in a solution

Short Answer

Refer to the pH values and the bicarbonate / carbonic acid buffer formula to answer the following questions.

pH Values

pH 9   baking soda

pH 7   water

pH 3   vinegar

9. Is vinegar more acidic or more alkaline than water?

10. Does baking soda have a higher concentration of H^+ or OH^- than water?

11. Are the concentrations of H^+ and OH^- equal in baking soda, water, or vinegar?

12. Water has _____ times more H^+ than baking soda.

Bicarbonate / carbonic acid buffer formula

$$HCO_3^- + OH^- \leftrightarrow H_2CO_3$$

13. Which molecule accepts H^+ when the blood becomes too acidic?

14. Which molecule donates H^+ when the blood becomes too alkaline?

15. Which molecule will increase in concentration when the blood becomes too acidic?

16. Which molecule will increase in concentration when the blood becomes too alkaline?

2.5   The organic molecules of living organisms

2.6   Carbohydrates: Used for energy and structural support

2.7   Lipids: Insoluble in water

Matching

Select from the following list the correct term for each definition below.

organic molecule	**hydrolysis**	**glycogen**
dehydration synthesis	**monosaccharide**	**starch**
oligosaccharide	**cellulose**	**lipid**
polysaccharide	**triglyceride**	**steroid**
phospholipid	**cholesterol**	**fatty acid**

_____ 1. a long straight or branched chain containing thousands of monosaccharides

_____ 2. a polysaccharide used by plants for structural support

_____ 3. the class of biochemical compounds that contains triglycerides, phospholipids, and steroids

_____ 4. a group of molecules classified as lipids because they are insoluble in water

_____ 5. a molecule that contains carbon covalently bonded to other elements

_____ 6. a steroid that forms part of the cell membrane

_____ 7. a reaction in which small molecules are covalently bonded by the removal of a water molecule

_____ 8. a lipid composed of one molecule of glycerol and three fatty acids

_____ 9. a reaction in which macromolecules are broken apart by the addition of a water molecule

_____ 10. a lipid that forms an important part of the cell membrane

_____ 11. a short string of monosaccharides linked together

_____ 12. the storage polysaccharide found in plants

_____ 13. the simplest carbohydrate, consisting of a single sugar unit

_____ 14. the storage polysaccharide found in animals

_____ 15. a chain of hydrocarbons with a terminal carboxyl group

Fill-in-the-Blank

Referenced sections are indicated in parentheses.

16. _____ are large organic molecules consisting of thousands of smaller molecules. (2.5)

17. The synthesis of macromolecules from smaller molecules _____ energy. (2.5)

18. The breakdown of macromolecules _____ energy. (2.5)

19. The four classes of organic molecules made by living organisms are _____, _____, _____, and _____. (2.5)

20. _____ is a six-carbon monosaccharide that serves as an important source of energy for cells. (2.6)

Labeling

Use the words below to identify each molecule or molecular region indicated in Figure 2.1.

a. glycerol

b. steroid

c. monosaccharide

d. fatty acids

e. unsaturated fatty acid

f. triglyceride

g. disaccharide

h. polar head

i. saturated fatty acid

j. phospholipid

k. fatty-acid tails

21. _____

22. _____

23. _____

_____ 27.

24. _____

25. _____   26. _____

_____ 28.

_____ 29.

30. _____

31. _____

Figure 2.1

2.8 Proteins: Complex structures constructed of amino acids

2.9 Nucleic acids store genetic information

2.10 ATP carries energy

Matching

Select from the following list the correct term for each definition below.

denaturation	amino acids	protein	catalyst
polypeptide	DNA, RNA	enzyme	ADP
nucleotides	ATP		

_____ 1. a molecule produced by the breakdown of ATP

_____ 2. single units joined by covalent bonds in DNA and/or RNA

_____ 3. a protein that speeds up a chemical reaction

_____ 4. a nucleotide that functions as the universal energy source for cells

_____ 5. single units joined by covalent bonds to form a protein

_____ 6. a chain of more than 100 amino acids

_____ 7. permanent damage to protein structure resulting in loss of function

_____ 8. a string of 3–100 amino acids

_____ 9. nucleic acids

_____ 10. a substance that speeds up the rate of a chemical reaction

True/False

Mark each statement true or false. If a statement is marked false, replace the underlined term or phrase with the term or phrase needed to make the statement true. Referenced sections are indicated in parentheses.

_____ 11. An enzyme can facilitate a reaction in which two reactants are joined to form a substrate. (2.8)

_____ 12. A nucleotide consists of a sugar, a phosphate, and a base. (2.9)

_____ 13. RNA can aid in the regulation of DNA. (2.9)

_____ 14. RNA contains the instructions for producing protein. (2.9)

_____ 15. The four bases found in DNA nucleotides are adenine, thymine, guanine, and uracil. (2.9)

_____ 16. RNA is a single-stranded molecule containing a deoxyribose sugar. (2.9)

_____ 17. The breakdown of ATP produces ADP, P, and energy. (2.10)

Short Answer

18. What is the monomer of a protein?

19. What is attached to each single bond surrounding the central carbon atom of an amino acid?

20. When is a polypeptide referred to as a protein?

21. List and describe the four levels of protein structure.

22. What causes denaturation of a protein?

23. How is enzyme shape an example of the importance of homeostasis?

Figure 2.2

Labeling

Use the terms below to identify each molecule or molecular region indicated in Figure 2.2.

a. triphosphate   e. ribose

b. amino group   f. carboxylic acid group

c. R group   g. ATP

d. adenine   h. amino acid

Looking Back/Looking Forward

1. Chapter 1, Section 1.1, describes the molecules of life found in all living organisms. How are these molecules different from those found in nonliving things?

2. Chapter 3 discusses the importance of phospholipids in the structure of the plasma membrane, the membrane that surrounds all cells. Cells are bathed in a watery fluid, both internally and externally. What characteristics of a phospholipid make it ideally suited to function as part of the plasma membrane?

Chapter Test

Multiple Choice

1. Which of the following statements about matter is *false*?
   a. Matter is anything that has mass and occupies space.
   b. Matter is composed of elements.
   c. An element is a pure form of matter that cannot be broken down to a simpler form.
   d. All elements are composed of identical atoms.

2. The atomic number of nitrogen is 7. How many electrons will there be in the 3rd shell of a nitrogen atom?
   a. 7
   b. 5
   c. 2
   d. 0
   e. insufficient information to answer the question

3. The structural formula O=O indicates that two atoms of oxygen are joined by a _____ bond.
   a. covalent
   b. ionic
   c. hydrogen
   d. double
   e. a and d
   f. b and d

4. Atoms are electrically neutral when they have equal numbers of
   a. protons and neutrons.
   b. neutrons and electrons.
   c. protons and electrons.
   d. protons, neutrons, and electrons.

5. Sodium has an atomic number of 11. This indicates that sodium has
   a. 11 neutrons.
   b. a total of 11 particles in the nucleus.
   c. 11 protons.
   d. a total of 11 particles in the entire atom.

6. Most atoms of hydrogen have no neutrons. An atom of hydrogen containing one neutron would
   a. be an ion.
   b. be an isotope.
   c. be a molecule.
   d. also have an additional proton.

7. One atom of C-12 has an atomic number of 6 and a mass number of 12. Which of the following is a true statement about C-14?
   a. C-14 has fewer protons than C-12.
   b. C-14 has more neutrons than C-12.
   c. C-12 has greater mass than C-14.
   d. C-12 is neutral, while C-14 carries a charge.

8. Atoms are most stable when
   a. they have the same number of protons as neutrons.
   b. they have the same number of protons as electrons.
   c. their outermost occupied electron shell is partially filled.
   d. their outermost occupied electron shell is full.

9. Covalent bonds involve _____ electrons, and ionic bonds involve _____ electrons.
   a. transferring, sharing
   b. transferring, transferring
   c. sharing, transferring
   d. sharing, sharing

10. Radioisotopes are *not* used to
    a. trace the movement of molecules in the body.
    b. trigger the movement of electrons between orbitals.
    c. destroy some types of cancer cells.
    d. determine the age of some rocks and fossils.

11. Cl^- is created when Cl:
    a. gains one electron
    b. loses one proton
    c. shares an electron with another atom
    d. forms hydrogen bonds with a water molecule

12. Which of the following is a *False* statement about energy?
    a. Energy is required to join atoms and form molecules, but not to break apart atoms within a molecule.
    b. Chemical bonds contain potential energy.
    c. Work is performed using kinetic energy.
    d. Electrons in the third shell of an atom have more potential energy than electrons in the second shell.

13. The covalent bonds between oxygen and hydrogen in a water molecule
    a. cannot be broken.
    b. can be broken, producing two hydrogen atoms and one oxygen atom.
    c. can be broken, producing $2 H^+$ and $1 O^-$.
    d. can be broken, producing $1 H^+$ and $1 OH^-$.

14. Which of the following occurs in a hydrogen bond?
    a. A positively charged hydrogen atom in one water molecule is attracted to a negatively charged oxygen atom in another water molecule.
    b. The electrons in the hydrogen bond spend more time orbiting the oxygen atom than the hydrogen atom.
    c. The electrons in the bond spend more time orbiting the hydrogen atom than the oxygen atom.
    d. a and b
    e. a and c

15. An acid is a molecule that _____ H^+, while a base is a molecule that _____ H^+.
    a. absorbs, releases
    b. releases, absorbs
    c. reacts with, does not react with
    d. decreases, increases

16. Blood pH is maintained by the carbonic acid/bicarbonate buffer system: $HCO_3^- + H^+ \leftrightarrow H_2CO_3$. If the blood becomes too acidic, which of the following will occur?
    a. Carbonic acid will release H^+.
    b. Carbonic acid will absorb H^+.
    c. Bicarbonate will release H^+.
    d. Bicarbonate will absorb H^+.

17. The synthesis of macromolecules from smaller molecules
    a. releases energy.
    b. requires energy.
    c. is accomplished by hydrolysis.
    d. requires the addition of water.

18. Monosaccharides
    a. contain C, H, and O in a 1–2–1 ratio.
    b. contain C and H in a 1–1 ratio.
    c. contain equal amounts of C and H, with a small amount of O.
    d. contain equal amounts of C and O, with a small amount of H.

19. Which of the following is a carbohydrate with structural importance in plants?
    a. glycogen
    b. glucose
    c. starch
    d. cellulose

20. Triglycerides with a full complement of two H atoms for each C atom in their fatty-acid tails are
    a. phospholipids.
    b. steroids.
    c. saturated.
    d. unsaturated.

21. Proteins function as catalysts when they
    a. act as enzymes.
    b. speed up the rate of a chemical reaction.
    c. store genetic information.
    d. a and b

22. The three-dimensional protein shape maintained by hydrogen and disulfide bonds is the
    a. primary structure.
    b. secondary structure.
    c. tertiary structure.
    d. quaternary structure.

23. What molecule is composed of a sugar, a phosphate, and a base?
    a. an amino acid
    b. a nucleotide
    c. a quaternary protein
    d. a saturated fat

24. Which of the following is *not* different in RNA and DNA?
    a. the phosphate group
    b. the sugar
    c. the presence of uracil
    d. the number of strands

25. The breakdown of ATP produces
    a. ADP.
    b. free phosphate.
    c. energy.
    d. all of the above.

Self Test

Each of the Key Concepts listed at the beginning of this chapter has been rewritten as a question below. After successfully completing the study guide exercises and the Chapter Test, you should be able to answer each of these questions. Refer to the Key Concepts list at the beginning of this chapter to check your answers.

1. What is matter composed of? What are elements composed of? What are atoms composed of?

2. What is indicated by the atomic number? What is indicated by the mass number?

3. What is an isotope?

4. What is energy? What type of energy is possessed by electrons based on their position?

5. What atoms tend to form chemical bonds?

6. Name three main types of bonds and describe how each is formed.

7. What characteristics of water make it a good biological solvent?

8. How does water help to regulate body temperature?

9. What is an acid? What is a base?

10. What is the pH scale?

11. What is the function of a buffer?

12. How can organic molecules be described?

13. How does dehydration synthesis differ from hydrolysis?

14. What are the four classes of organic molecules synthesized by living organisms?

15. Simple sugars and polysaccharides are both involved in energy; how are their roles different?

16. What are the three types of lipids and their functions?

17. What are proteins composed of? How are proteins characterized?

18. What is an enzyme?

19. What are nucleic acids composed of?

20. What are the functions of DNA, RNA, and ATP?

3

Structure and Function of Cells

Chapter Summary and Key Concepts

After reading and studying this chapter you should know the following:

Sections 3.1, 3.2, 3.3, 3.4

1. The cell doctrine states that all living things are composed of cells, a cell is the smallest unit of life, and that all cells come from preexisting cells.

2. Eukaryotic cells have a membrane-bound nucleus; prokaryotic cells do not.

3. Eukaryotic cells have a plasma membrane, a nucleus, and cytoplasm.

4. Cells remain small to stay efficient.

5. The plasma membrane is selectively permeable and consists of a lipid bilayer, proteins, cholesterol, and carbohydrates that extend from the outer surface.

6. Passive transport moves molecules with the gradient and requires no energy; active transport moves molecules against the gradient and requires energy.

7. Membrane proteins are required for some processes of diffusion, facilitated diffusion, and active transport.

8. Osmosis is the net diffusion of water across a selectively permeable membrane.

9. Endocytosis and exocytosis move materials in bulk. Endocytosis moves material into the cell, and exocytosis moves material out of the cell.

10. Receptor proteins receive and transmit information across the plasma membrane.

11. The sodium-potassium pump is an important active transport system that helps to maintain ion gradients and cell volume.

Sections 3.5, 3.6

12. The nucleus is surrounded by a nuclear membrane consisting of two lipid bilayers studded with pores.

13. Ribosomes synthesize proteins and may be free in the cytoplasm or bound to the rough endoplasmic reticulum.

14. The smooth endoplasmic reticulum synthesizes nonprotein molecules and packages proteins in vesicles for movement to the Golgi apparatus.

15. The Golgi apparatus refines proteins and packages them to be shipped.

16. Vesicles are membrane-bound sacs that enclose substances within the cell. They may be secretory, endocytotic, for shipping and storage, or lysosomal.

17. Mitochondria harvest the energy from ingested nutrients and use it to synthesize ATP.

18. Cell structures that contain microtubules include the cytoskeleton, which provides structural support; cilia and flagella that aid movement; and centrioles that participate in cell division.

Section 3.7

19. Cells use and transform matter and energy through chemical reactions. Metabolism is the sum of all chemical reactions in an organism.

20. Cellular respiration breaks down glucose in the presence of oxygen to yield 36 molecules of ATP. In the absence of oxygen, a cell can produce two molecules of ATP using anaerobic pathways.

Exercises

Complete the exercises for each section after you have read and studied the section. If you cannot answer some questions, or if you answer them incorrectly, return to the chapter and review this information. You may find it helpful to work on only one section at a time. When you have completed all sections, take the Chapter Test as an indicator of your mastery of this topic.

3.1   Cells are classified according to their internal organization

3.2   Cell structure reflects cell function

3.3   A plasma membrane surrounds the cell

3.4   Molecules cross the plasma membrane in several ways

Matching

Select from the following list the correct term for each definition below.

cell doctrine	nucleus	cytoplasm
plasma membrane	passive transport	eukaryotes
organelles	prokaryotes	selective permeability
diffusion	osmosis	facilitated transport
transport protein	exocytosis	endocytosis

receptor protein   sodium-potassium pump   vesicle

hypertonic solution   hypotonic solution   isotonic solution

active transport

_____ 1. the movement of molecules from one area to another as a result of random motion

_____ 2. material within a cell, composed of fluid and organelles

_____ 3. a solution with a concentration of solutes higher than the intra-cellular fluid

_____ 4. diffusion of substances through the plasma membrane with the help of transport proteins

_____ 5. three general principles that describe cells and their relationship to living things

_____ 6. movement of a substance through the plasma membrane against the concentration gradient

_____ 7. movement of material out of the cell by fusion of a vesicle with the plasma membrane

_____ 8. the outer membrane that surrounds all cells

_____ 9. the nature of the plasma membrane to allow some substances to cross by diffusion, but not others

_____ 10. an active-transport pump important in maintaining cell volume

_____ 11. movement of materials into the cell by enclosure in a vesicle formed by the plasma membrane

_____ 12. membrane proteins with specific receptor sites that can trigger biochemical events within the cell

_____ 13. organisms composed of cells that have a membrane-bound nucleus

_____ 14. a solution with the same solute concentration as the intracellular fluid

_____ 15. the net diffusion of water across a selectively permeable membrane

_____ 16. transport of a molecule through the plasma membrane without the use of energy

_____ 17. organisms whose cells do not have a membrane-bound nucleus

_____ 18. a solution with a concentration of solutes lower than the intracellular fluid

_____ 19. structure within a cell that carries out specialized functions

_____ 20. membrane-bound structure that contains a cell's genetic material

_____ 21. membrane-bound structure that transports substances into, out of, and within a cell

_____ 22. required for some membrane transport processes to move a substance across the plasma membrane

Fill-in-the-Blank

Referenced sections are indicated in parentheses.

23. Eukaryotic cells contain three basic structural components, the _____,
_____, and _____ _____. (3.1)

24. Some cells increase their surface area without significantly increasing their volume with a
projection of the cell membrane called _____. (3.2)

25. The _____ _____ microscope bombards
a sample with a beam of electrons to produce a two-dimensional image. (3.2)

26. The plasma membrane is composed of two layers of phospholipids called the
_____ _____. (3.3)

27. The changing structure and protein pattern of the plasma membrane is referred to by the
term _____ _____. (3.3)

28. A _____ exists when there is a difference in the concentration of a solution
in different areas. (3.4)

29. _____ exists when molecules are diffusing randomly, but equally, in all
directions. (3.4)

30. The fluid pressure required to oppose osmosis is called _____
_____. (3.4)

31. When ATP is used to power active transport, it will break down to produce
_____, _____, and _____. (3.4)

32. The area outside a cell is referred to as the _____ environment. (3.4)

33. When the sodium-potassium pump is being used to reduce cell volume, the
_____ of the pump will increase. (3.4)

Diagram

34. Draw a section of a plasma membrane, labeling each component. You
should include the following components in your diagram: (a) phospholipid,
(b) phospholipid bilayer, (c) cholesterol, (d) glycoprotein, (e) receptor protein,
(f) transport protein, (g) open protein channel, (h) gated protein,
(i) carbohydrate groups, (j) cytoskeleton filaments.

Completion

*For each of the membrane transport processes listed below, indicate in the table **all** of the following characteristics that apply (characteristics may be used more than once).*

a. requires energy

b. does not require energy

c. moves a substance through the lipid bilayer of the plasma membrane

d. moves a substance through a protein channel

e. moves a substance with the help of a transport protein

f. moves a substance with the gradient

g. moves a substance against the gradient

h. moves a substance within a vesicle

i. moves water only

Transport Process	Characteristics
35. Diffusion through the lipid bilayer	
36. Diffusion through protein channels	
37. Facilitated transport	
38. Osmosis	
39. Active transport	
40. Endocytosis	
41. Exocytosis	

Word Choice

Circle the accurate term or phrase to complete each sentence.

42. Cells must maintain a (small/large) size to be efficient. (3.2)

43. The concentration of water in a solution is (the same as/opposite to) the concentration of solutes in the solution. (3.4)

44. Polar molecules (can/cannot) cross the lipid bilayer portion of the cell membrane. (3.4)

45. When a substance moves against the gradient, it moves from an area of (low/high) concentration to an area of (low/high) concentration. (3.4)

46. Gated protein channels are normally (open/closed) and will (open/close) under certain conditions. (3.4)

47. Moving a substance with the gradient (does/does not) require energy. (3.4)

48. The sodium-potassium pump moves three sodium ions (into/out of) the cell and two potassium ions (into/out of) the cell. (3.4)

49. In a hypertonic environment, water will move (into/out of) the cell. (3.4)

Critical Thinking

50. The presence of cholesterol in the cell membrane helps to maintain membrane fluidity by preventing the membrane from becoming too fluid at high temperatures and too solid at cooler temperatures. Consider and describe the properties of saturated and unsaturated fats in proposing a mechanism for this action of cholesterol.

3.5   Internal structures carry out specific functions

3.6   Cells have structures for support and movement

Crossword Puzzle

Across

1. Synthesizes ATP
4. Vesicles containing enzymes that detoxify wastes
9. Rodlike structures composed of microtubules that aid the process of cell division
10. An organelle that contains the cell's genetic material
12. The _____ endoplasmic reticulum synthesizes lipids
14. The endoplasmic _____ is a membranous tunnel-like system running through the cytoplasm

Down

2. A network of microtubules and microfilaments that provides structural support to the cell
3. The _____ apparatus refines proteins and packages them to be shipped to their final destination
5. Responsible for protein synthesis in the cytoplasm
6. Vesicles containing powerful digestive enzymes
7. The _____ endoplasmic reticulum has attached ribosomes
8. Long whiplike projection, found in humans only on sperm cells
11. A membrane-bound sphere that encloses something within the cell
13. Short hairlike projections found covering a cell surface

Labeling

Label the indicated structures in Figure 3.1.

Figure 3.1

Critical Thinking

28. Consider the movement of a secretory vesicle from the Golgi apparatus to the plasma membrane where the contents of the vesicle will be released. Such vesicles often carry receptors that will become cell surface receptors on the plasma membrane. Are these receptors found on the internal or external surface of the secretory vesicle membrane? Explain your answer.

3.7 Cells use and transform matter and energy

True/False

The following statements refer to the pathway: $A \rightarrow B \rightarrow C$

Mark each statement as true or false. If you mark a statement as false, replace the underlined term or phrase with the term or phrase needed to make the statement true.

_____ 1. This is a <u>linear</u> pathway.

_____ 2. If C is a larger molecule than A, this is a <u>catabolic</u> pathway.

_____ 3. This pathway probably requires <u>two</u> enzymes.

_____ 4. In this pathway, B is the <u>substrate</u> in the reaction $A \rightarrow B$.

_____ 5. In this pathway, B is the <u>product</u> in the reaction $B \rightarrow C$.

Short Answer

Answer the following questions about each stage of Cellular Respiration.

Glycolysis:

6. What is the starting molecule?

7. Where in the cell does glycolysis occur?

8. What is the end product of the energy investment steps of glycolysis?

9. What is the end product of the energy-yielding steps of glycolysis?

10. NAD^+ accepts one H^+ and two electrons to become what molecule?

11. What is the function of NADH?

Preparatory Step:

12. Where in the cell does the preparatory step occur?

13. What is the end product?

14. Is CO_2 released?

Citric Acid Cycle:

15. Where in the cell does the citric acid cycle occur?

16. Is the citric acid cycle a linear or cyclic pathway?

17. What energy-carrying coenzymes harvest high energy H^+ and electrons during the cycle?

18. Does the citric acid cycle complete the catabolism of glucose?

19. Only 4 ATP molecules have been generated by the end of the citric acid cycle; where is the remaining energy that has been harvested from the glucose molecule?

Electron Transport System:

20. Where in the cell does the electron transport system occur?

21. As electrons are passed from one protein carrier molecule to the next, is energy required or released?

22. Why do carrier proteins need energy to transport H^+ from the inner compartment to the outer compartment of the mitochondria?

23. What molecule serves as a channel for the diffusion of H^+ from the outer compartment to the inner compartment of the mitochondria?

24. What provides the energy for the synthesis of ATP by ATP synthase?

25. What term refers to the synthesis of ATP using the energy source referred to in #24?

Critical Thinking

26. Why do fats store more than twice the energy of carbohydrates?

27. What result might you expect if the energy released by electron transport at the cristae was not coupled to ATP production in mitochondria?

Looking Back/Looking Forward

1. Chapter 2 discussed organic macromolecules important in living things. Is the synthesis of a protein from individual amino acids anabolic or catabolic? Does it require energy, or is energy released? Explain your answers.

2. Chapter 14 will discuss the digestive system and the absorption of nutrients. During this process, proteins are broken down into amino acids that are transported in intestinal cells that already have a high concentration of amino acids. What transport process will be responsible for this movement of amino acids?

Chapter Test

Multiple Choice

1. Which of the following is not part of the cell doctrine?
   a. All living things are composed of cells.
   b. Cells must remain small to function effectively.
   c. Cells are the smallest units of life.
   d. All cells come from preexisting cells.

2. You suspect a cell to be prokaryotic. Which of the following would indicate you are mistaken?
   a. The cell has a plasma membrane.
   b. The cell has a nuclear membrane.
   c. The cell has genetic material.
   d. The cell has cytoplasm.

3. The plasma membrane is composed of lipids arranged in a
   a. single layer with the hydrophilic heads near the extracellular fluid.
   b. single layer with the hydrophobic tails near the extracellular fluid.
   c. bilayer with the tails facing the intracellular and extracellular fluid.
   d. bilayer with the heads facing the intracellular and extracellular fluid.

4. Cholesterol in the plasma membrane functions
   a. as a receptor molecule.
   b. as a transport molecule.
   c. to make the membrane more rigid.
   d. to make the membrane more permeable.

5. Cells with microvilli
   a. have less surface area than other cells.
   b. have a surface covered with small projections of the plasma membrane.
   c. have multiple nuclei.
   d. are associated with transmitting information between cells.
   e. b and d.

6. The "fluid mosaic" model of the plasma membrane refers to
   a. the ability of proteins to drift about in the membrane.
   b. the fluidity of the lipid bilayer.
   c. the presence of organelles in the cytoplasm.
   d. a and b.
   e. all of the above.

7. Passive transport
   a. requires energy.
   b. relies on the process of diffusion.
   c. is dependent on active transport.
   d. moves a substance against the gradient.

8. Which of the following moves bulk substances into the cell by forming a membrane-bound vesicle?
   a. passive transport
   b. osmosis
   c. endocytosis
   d. exocytosis

9. Oxygen and carbon dioxide move into and out of the cell by
   a. diffusion through the lipid bilayer.
   b. facilitated diffusion.
   c. active transport.
   d. osmosis.

10. Substance X is more concentrated in the intracellular fluid than the extracellular fluid. Substance X will move into the cell by
    a. diffusion.
    b. facilitated diffusion.
    c. exocytosis.
    d. active transport.

11. How does facilitated diffusion differ from active transport?
    a. Facilitated diffusion requires transport proteins.
    b. Facilitated diffusion does not require energy.
    c. Facilitated diffusion moves a substance down the concentration gradient.
    d. Both b and c.
    e. All of the above.

12. The sodium-potassium pump is important in maintaining cell volume because
    a. it transports water into and out of the cell.
    b. it blocks the movement of water with membrane proteins.
    c. it transports ions, which results in the movement of water by osmosis.
    d. it counteracts the effect of osmotic pressure.

13. Which of the following events may contribute to a transfer of information across the plasma membrane?
    a. Receptor proteins receive and transmit information.
    b. A molecule binds a receptor site on a receptor protein.
    c. Endocytosis brings receptor proteins into the cell.
    d. Both a and b.
    e. All of the above.

14. A cell is placed in a hypotonic environment. Which of the following will occur?
    a. Water will move into the cell and the cell will swell.
    b. Water will move out of the cell and the cell will shrink.
    c. Solutes will move into the cell and the cell will shrink.
    d. Solutes will move out of the cell and the cell will swell.

15. The nucleolus is
    a. an area of concentrated DNA.
    b. an area of RNA synthesis.
    c. an area outside the nucleus where ribosomes are assembled.
    d. a small additional nucleus found in large cells.

16. A cell that lacked smooth endoplasmic reticulum (ER) would be unable to
    a. synthesize lipids.
    b. detoxify wastes.
    c. manufacture ATP.
    d. synthesize proteins for use outside the cell.

17. Vesicles fusing with the Golgi apparatus would have arisen from the
    a. plasma membrane.
    b. smooth ER.
    c. mitochondria.
    d. nucleus.

18. Vesicles fusing with the plasma membrane in the process of
    _____ would have arisen from the
    _____.
    a. exocytosis, Golgi apparatus
    b. endocytosis, Golgi apparatus
    c. exocytosis, rough ER
    d. endocytosis, nucleus

19. Cellular debris will be removed by
    a. peroxisomes.
    b. endocytotic vesicles.
    c. secretory vesicles.
    d. lysosomes.

20. Which of the following is composed of a circle of nine microtubule pairs
    surrounding a single pair in the center?
    a. centrioles
    b. cilia
    c. mitochondria
    d. ribosomes

21. Glycolysis is all of the following *except*
    a. aerobic.
    b. a linear pathway.
    c. a process that occurs in the cytoplasm.
    d. a process that produces ATP.

22. The breakdown of macromolecules is a(n) _____ process
    and _____ energy.
    a. anabolic, requires
    b. anabolic, releases
    c. catabolic, requires
    d. catabolic, releases

23. Which of the following stages of cellular respiration does not contain any
    carbon atoms from the original glucose molecule?
    a. citric acid cycle
    b. electron transport system
    c. glycolysis
    d. formation of acetyl CoA

24. How much ATP could a cell produce if no molecules of NAD⁺ and FAD were available?
    a. 0
    b. 4
    c. 12
    d. 32

25. What would happen to ATP production in the mitochondria if the inner mitochondrial membrane was freely permeable to H⁺?
    a. ATP synthesis would proceed with no change.
    b. ATP synthesis would be slightly reduced.
    c. ATP synthesis would be slightly increased.
    d. ATP synthesis would stop.

Self Test

Each of the Key Concepts listed at the beginning of this chapter has been rewritten as a question below. After successfully completing the study guide exercises and the Chapter Test, you should be able to answer each of these questions. Refer to the Key Concepts list at the beginning of this chapter to check your answers.

1. What three principles are stated by the cell doctrine?

2. How are eukaryotic cells different from prokaryotic cells?

3. What three basic components are found in eukaryotic cells?

4. Why are all cells small?

5. How permeable is the plasma membrane? What components are found as part of the plasma membrane?

6. How does passive transport differ from active transport?

7. What transport processes require membrane proteins?

8. What is osmosis?

9. What occurs in endocytosis and exocytosis?

10. What is the role of receptor proteins?

11. Why is the sodium-potassium pump important to the cell?

12. How would you describe the structure of the nuclear membrane?

13. What is the function of ribosomes? Where are ribosomes located?

14. What is the function of the smooth endoplasmic reticulum?

15. What is the function of the Golgi apparatus?

16. What are vesicles? List four different types of vesicles.

17. What is the function of mitochondria?

18. List four cell structures that contain microtubules, and describe the function of each.

19. How do cells use and transform energy? What is metabolism?

20. What molecules are involved in the process of cellular respiration? How much ATP is produced? How is this process different in the absence of oxygen?

4

From Cells to Organ Systems

Chapter Summary and Key Concepts

After reading and studying this chapter you should know the following:

Sections 4.1, 4.2, 4.3

1. Multicellular organisms consist of many cells, allowing them to achieve greater size and greater ability to maintain a desired environment.

2. Tissues are groups of cells that perform common functions. The four major tissue types of the body are epithelial, connective, muscle, and nervous tissue.

3. Epithelial tissue covers or lines body surfaces or cavities, and is classified by the number of cell layers and cell shape.

4. Glands are a type of epithelial tissue specialized to secrete a product.

5. A basement membrane anchors epithelial tissue to the underlying connective tissue. Junctions anchor epithelial cells to other cells.

6. Connective tissue supports soft organs and connects parts of the body.

7. Connective tissue consists of a nonliving *matrix* of background substance and fibers, and cells responsible for secreting the matrix.

8. Connective tissue can be classified as fibrous or specialized.

9. Fibrous connective tissue includes the following types: loose, dense, elastic, and reticular.

10. Specialized connective tissue includes the following types: cartilage, bone, blood, and adipose tissue.

11. Each type of connective tissue will vary in the type of matrix, cells, and fibers it contains. The structure of the tissue allows it to perform its function.

Sections 4.4, 4.5

12. Muscle tissue consists of cells specialized to produce movement. Muscle tissue types include skeletal, smooth, and cardiac.

13. Skeletal muscle cells are voluntary, multinucleated, and attached to bones.

14. Smooth muscle cells are involuntary, have a single nucleus, and are found surrounding hollow organs and tubes.

15. Cardiac muscle cells are involuntary, have a single nucleus, and are found in the heart.

16. Nervous tissue consists of cells specialized to generate and transmit electrical impulses used for communication in the body. Nervous tissue cells include neurons responsible for communication, and glial cells that support the neurons.

Sections 4.6, 4.7, 4.8

17. Organ systems are composed of groups of organs that perform a broad function. The human body has 11 organ systems.

18. Tissue membranes line body cavities. The four major types of tissue membranes are serous, mucous, synovial, and cutaneous.

19. The integumentary system consists of the epidermis and the dermis, and it has a role in protection from dehydration, injury, and invasion; regulation of body temperature; synthesis; and sensation.

20. Multicellular organisms maintain homeostasis using negative feedback control systems. These systems include a controlled variable, a sensor, a control center, and an effector.

Exercises

Complete the exercises for each section after you have read and studied the section. If you cannot answer some questions, or if you answer them incorrectly, return to the chapter and review this information. You may find it helpful to work on only one section at a time. When you have completed all sections, take the Chapter Test as an indicator of your mastery of this topic.

4.1   Tissues are groups of cells with a common function

4.2   Epithelial tissues cover body cavities and surfaces

4.3   Connective tissue supports and connects body parts

Matching

Select from the following list the correct term for each definition below.

tissue   epithelial tissue   exocrine gland

glands   fibroblasts   endocrine gland

basement membrane   bone   adipocytes

connective tissue   reticular fibers   blood

elastic fibers   collagen fibers   adipose tissue

cartilage

_____ 1. a connective tissue composed primarily of collagen fibers, with chondroblasts in lacunae

_____ 2. fat cells

_____ 3. connective tissue protein fibers that allow a tissue to stretch

_____ 4. a connective tissue specialized for fat storage

_____ 5. a gland that secretes its product into a hollow organ or duct

_____ 6. a group of specialized cells that perform a common function

_____ 7. a connective tissue consisting of cells suspended in a fluid called plasma

_____ 8. sheets of cells that line or cover body surfaces or cavities

_____ 9. connective tissue cells that secrete protein fibers

_____ 10. a gland that secretes hormones into the bloodstream

_____ 11. a tissue that supports and connects parts of the body

_____ 12. epithelial tissue specialized to secrete a product

_____ 13. connective tissue fibers that support or provide an internal framework for a body organ

_____ 14. connective tissue protein fibers that confer strength to a tissue

_____ 15. a supporting layer beneath epithelial tissue that anchors it to connective tissue

_____ 16. a hard connective tissue containing calcium and phosphate

Labeling

Label each epithelial tissue in Figure 4.1 with (a) tissue type and (b) function.

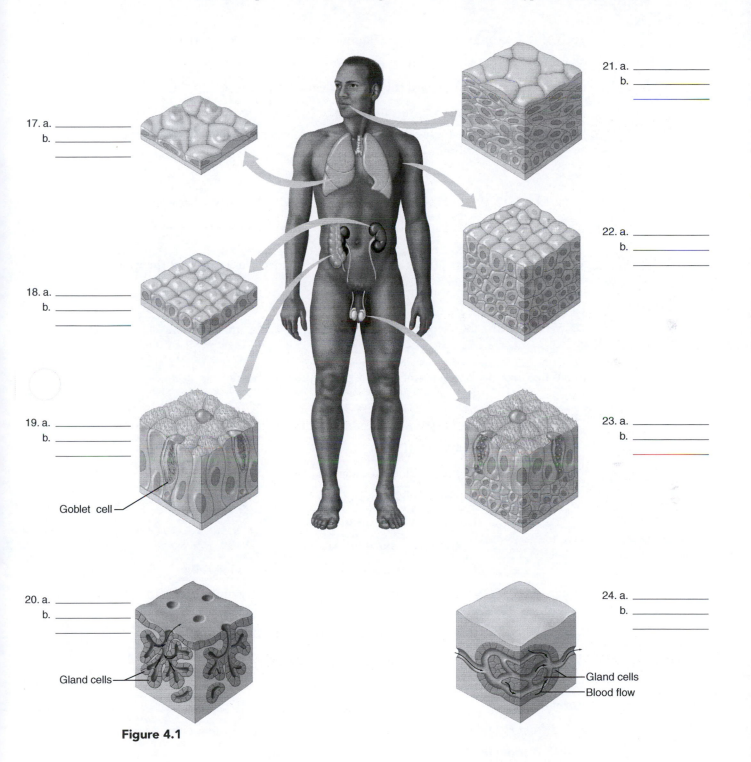

17. a. _____
    b. _____
       _____

18. a. _____
    b. _____
       _____

19. a. _____
    b. _____
       _____

    Goblet cell

20. a. _____
    b. _____
       _____

    Gland cells

21. a. _____
    b. _____
       _____

22. a. _____
    b. _____
       _____

23. a. _____
    b. _____
       _____

24. a. _____
    b. _____
       _____

    Gland cells
    Blood flow

Figure 4.1

Short Answer

25. List four functions performed by connective tissues.

26. What are the two components found in most types of connective tissue?

27. What are the two general categories of connective tissue?

28. a. What are the three general components found in the extracellular matrix of all types of fibrous connective tissue?

    b. List the three types of fibers found in fibrous connective tissue, and the characteristics of each fiber.

    c. List the four cell types found in fibrous connective tissue, and the function of each.

29. What are the four types of fibrous connective tissue, and where is each found?

30. What are the four types of specialized connective tissue?

31. Identify the type of specialized connective tissue in each description below:

    _____ a. has a fluid matrix

    _____ b. has no blood vessels

    _____ c. reduces friction in the joints

    _____ d. has very little ground substance

    _____ e. has a very hard matrix

    _____ f. has mostly matrix with only a few living cells

    _____ g. most of this tissue is located directly under the skin

4.4   Muscle tissues contract to produce movement

4.5   Nervous tissue transmits impulses

Matching

Select from the following list the correct term for each definition below.

skeletal muscle tissue   cardiac muscle tissue

smooth muscle tissue   nervous tissue

muscle tissue   neurons

glial cells

_____ 1. cells of the nervous tissue that support neurons

_____ 2. muscle tissue that surrounds hollow organs

_____ 3. a tissue specialized for movement

_____ 4. nervous tissue cells that generate and transmit electrical impulses

_____ 5. muscle tissue connected to bone by tendons

_____ 6. muscle tissue found in the heart

_____ 7. a tissue specialized for generating and transmitting electrical impulses

Completion

Complete the table by filling in the name or characteristics of each type of muscle tissue.

	Tissue Type	Location	# Nuclei per Cell	Voluntary or Involuntary
8.	a)	b)	c)	Voluntary
9.	a)	Heart	b)	c)
10.	Smooth muscle tissue	a)	b)	c)

4.6 Organs and organ systems perform complex functions

True/False

Mark each statement as true or false. If you mark a statement as false, replace the underlined term or phrase with the term or phrase needed to make the statement true.

_____ 1. A structure composed of two or more tissue types joined together to perform a function is a(n) <u>organ system</u>.

_____ 2. A <u>basement</u> membrane consists of a layer of epithelial tissue and a layer of connective tissue.

_____ 3. <u>Membrane</u> is a general term for a thin layer that covers or surrounds something.

_____ 4. <u>Serous</u> membranes line thin cavities between bones in movable joints.

_____ 5. A <u>midsagittal</u> plane divides the body into top and bottom portions.

_____ 6. A structure closer to a point of reference is described as <u>distal</u>.

_____ 7. The thoracic and abdominal cavities are separated by the <u>pleural cavity</u>.

Matching

Select from the following list the correct organ system for each description.

integumentary	endocrine	lymphatic
skeletal	digestive	respiratory
muscular	circulatory	reproductive
nervous	urinary	

_____ 8. Responsible for gas exchange between the air and the blood.

_____ 9. Produces or resists movement.

_____ 10. Transports material through the body, helps maintain body temperature, aids body defenses.

_____ 11. Produces eggs and sperm.

_____ 12. Excretes waste, maintains volume and composition of body fluids.

_____ 13. Detects stimuli in the external and internal environment and coordinates a response.

_____ 14. Provides the structural framework for movement of the body.

_____ 15. Provides protection from injury, infection, and dehydration; receives sensory input.

_____ 16. Returns excess tissue fluid to the circulatory system and aids the body defense mechanisms.

_____ 17. Produces hormones that regulate many body functions.

_____ 18. Provides the body with water and nutrients.

4.7 The skin as an organ system

Crossword Puzzle

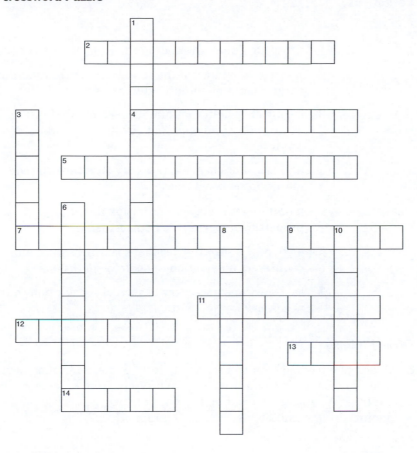

Across

2. Pigment-producing cells in the epidermis
4. Sweat glands produce sweat that contains this antibiotic secretion
5. The skin represents this organ system
7. A supporting layer for the skin that contains adipose tissue and loose connective tissue
9. Cells at the base of the epidermis that are constantly dividing
11. Small projections on the surface of the dermis
12. A waterproof protein produced by keratinocytes in the epidermis
13. The region of a hair that is found below the skin's surface
14. The region of a hair that extends above the skin's surface

Down

1. The outer layers of the epidermis are dead because these are not found in the epidermis
3. This type of muscle is found at the base of each hair follicle
6. The outer layer of the skin's epithelial tissue
8. A type of gland that produces oil
10. This type of epithelial cell is found in the epidermis

4.8   Multicellular organisms must maintain homeostasis

Definitions

Write the term(s) for each definition below.

_____ 1. the clear fluid surrounding the cells of a multicellular organism

_____ 2. relative constancy of the condition within the internal environment of a multicellular organism

_____ 3. a control system that maintains a desired internal environment by detecting and opposing deviations from the normal desired conditions

_____ 4. a physical or chemical property that must be controlled to maintain homeostasis

_____ 5. monitors and relays information about physical and chemical properties important to homeostasis

_____ 6. as part of a negative feedback control loop, sends signals to counteract a deviation in desired conditions

_____ 7. takes actions necessary to correct a deviation in conditions necessary for homeostasis

Critical Thinking

8. A positive feedback control system detects and strengthens deviations from the body's normal condition. How is this different from a negative feedback control system? Why would a positive feedback control system be ineffective in maintaining homeostasis with body temperature?

9. How does the composition of the interstitial fluid affect the movement of substances across the plasma membrane?

Looking Back/Looking Forward

1. Endocrine cells of the pancreas secrete the hormone insulin into the blood via the endomembrane system of each secretory pancreatic cell. What type of cellular transport mechanism is occurring?

2. Chapter 17 describes how cells reproduce and Chapter 18 discusses cancer, a disease that results from cell damage leading to uncontrolled cell division. Skin cancer arises when basal cells or melanocytes are damaged. Why are cells of the upper epidermis not a source of skin cancer?

Chapter Test

Multiple Choice

1. When a group of cells perform a common function they become a(n)
   a. organism.
   b. organ.
   c. tissue.
   d. organ system.

2. In looking at a cross section of kidney tubule under the microscope you notice a single layer of square-shaped cells lining the tubule. You identify these cells as _____ epithelial tissue.
   a. simple squamous
   b. simple cuboidal
   c. simple columnar
   d. stratified columnar

3. A simple epithelium would *not* function in
   a. transporting materials.
   b. absorbing nutrients.
   c. secreting wastes.
   d. providing protection.

4. All glands are involved in synthesis of a product, but only _____ glands produce hormones.
   a. salivary
   b. endocrine
   c. exocrine
   d. sweat

5. All of the following accurately describe a basement membrane *except*
   a. supportive noncellular layer
   b. functions to anchor cells to underlying tissue
   c. surrounds the plasma membrane of epithelial cells
   d. composed primarily of protein

6. Matrix is found in all of the following *except*
   a. adipose tissue.
   b. cartilage.
   c. elastic connective tissue.
   d. muscle.

7. Cardiac muscle cells must be able to communicate rapidly with each other, a process that involves the movement of ions from one cell to another. Which type of junction aids this process in cardiac muscle cells?
   a. tight junctions
   b. communication junctions
   c. adherence junctions
   d. gap junctions

8. A unique feature of connective tissue is that the cells of the tissue secrete the

   _____.

   a. plasma membrane
   b. junctions
   c. basement membrane
   d. matrix

9. Which of the following connective tissue elements are incorrectly matched with a characteristic?
   a. elastic fibers—allow tissue to stretch
   b. reticular fibers—secrete matrix
   c. collagen fibers—confer strength
   d. adipocytes—store fat

10. Loose connective tissue and dense connective tissue both contain collagen fibers. What is different about the collagen fibers in these two tissues?
    a. The chemical composition of the fibers.
    b. The orientation of the fibers.
    c. The collagen in dense connective tissue is more flexible than the collagen in loose connective tissue.
    d. The collagen fibers in dense connective tissue are repairable, while the collagen fibers in loose connective tissue are not.

11. Cartilage is slow to heal because
    a. it is a connective tissue.
    b. it is not specialized for growth.
    c. it is seldom injured.
    d. it has no blood vessels.

12. Gap junctions are found in
    a. skeletal muscle cells.
    b. smooth muscle cells.
    c. cardiac muscle cells.
    d. b and c.
    e. all of the above.

13. The walls of blood vessels contain _____ muscle that is

    _____.
    a. skeletal muscle, voluntary
    b. skeletal muscle, involuntary
    c. smooth muscle, voluntary
    d. smooth muscle, involuntary

14. In a neuron the _____ receives signals, and the

    _____ transmits signals.
    a. axon, cell body
    b. cell body, dendrite
    c. dendrite, axon
    d. axon, dendrite

15. The anterior cavity consists of
    a. the thoracic cavity and the spinal cavity.
    b. the cranial cavity and the thoracic cavity.
    c. the thoracic cavity, abdominal cavity, and pelvic cavity.
    d. the cranial cavity and the spinal cavity.

16. Which of the following tissue membranes are incorrectly matched with its location?
    a. cutaneous membranes—line the reproductive passages
    b. synovial membranes—line the joint cavity of moveable joints
    c. mucous membranes—line the airways and digestive tract
    d. serous membranes—line body cavities

17. Goblet cells are found in _____ and function to

    _____.
    a. connective tissue, attach tendons to bone
    b. columnar epithelial tissue, secrete mucus
    c. synovial membranes, lubricate body cavities
    d. the hypodermis, deliver nutrients to tissues

18. Which of the following systems produces hormones that regulate many body functions?
    a. integumentary
    b. nervous
    c. endocrine
    d. reproductive

19. Which of the following terms are incorrectly defined?
    a. proximal—farther away
    b. anterior—closer to the front
    c. transverse plane—divides the body into right and left sides
    d. a and c

20. The hypodermis
    a. is the middle layer of the skin.
    b. provides rigid support for the epidermis.
    c. contains fat cells.
    d. a and c
    e. b and c

21. Which of the following structures of the skin is incorrectly matched with a function or description?
    a. sebaceous glands—secrete an oily fluid
    b. follicle—surrounds the shaft of a hair
    c. hypodermis—contains fat cells
    d. keratinocytes—produce a waterproof protein

22. Which of the following is a false statement about interstitial fluid?
    a. The composition of the interstitial fluid should change regularly.
    b. Interstitial fluid is found between the cells of an organism.
    c. Interstitial fluid makes up the internal environment of an organism.
    d. Cells obtain nutrients from the interstitial fluid.

23. A drop in blood pressure results in constriction of the blood vessels. Blood pressure functions as the _____ in a negative feedback system.
    a. sensor
    b. effector
    c. control center
    d. controlled variable

24. As humans age the skin begins to sag and wrinkle. This is most likely due to a change in
    a. the cellular junctions.
    b. the blood supply to the skin.
    c. the type of collagen fibers produced.
    d. the amount of skeletal muscle present.

25. Contractions of the uterus during childbirth are an example of positive feedback because
    a. they are a short-term event.
    b. they are not required to maintain health.
    c. a change in a controlled variable is not monitored by sensors.
    d. a change in a controlled variable serves to amplify the change.

Self Test

Each of the Key Concepts listed at the beginning of this chapter has been rewritten as a question below. After successfully completing the study guide exercises and the Chapter Test, you should be able to answer each of these questions. Refer to the Key Concepts list at the beginning of this chapter to check your answers.

1. What are some advantages of being multicellular?

2. What is a tissue? What are the four major tissue types in the body?

3. Where are epithelial tissues found, and how are they classified?

4. What are glands? What is their function?

5. What is the function of a basement membrane? What is the function of a cellular junction?

6. What is the broad function of connective tissue?

7. What components make up connective tissue?

8. What are the two broad categories of connective tissue?

9. What are four types of fibrous connective tissue?

10. What are four types of specialized connective tissue?

11. How are connective tissue types different?

12. What is the function of muscle tissue? List three types of muscle tissue.

13. What are three characteristics of skeletal muscle cells?

14. What are three characteristics of smooth muscle cells?

15. What are three characteristics of cardiac muscle cells?

16. What is the function of nervous tissue? Name two types of nervous tissue cells and their function.

17. What is an organ system? How many organ systems are found in the human body?

18. What is the function of a tissue membrane? Name the four major types of tissue membranes.

19. What are the two main components of the integumentary system? What are the functions of this system?

20. How do multicellular organisms maintain homeostasis? What is involved in this mechanism?

5

The Skeletal System

Chapter Summary and Key Concepts

After reading and studying this chapter you should know the following:

Sections 5.1, 5.2, 5.3

1. The skeletal system consists of bones, ligaments, and cartilage.

2. The skeleton provides support, movement, and protection. In addition, the bones provide for formation of blood cells and mineral storage.

3. Compact bone tissue consists of osteons containing osteocytes in lacunae that communicate via canaliculi.

4. Spongy bone tissue is a latticework of trabeculae containing osteocytes in lacunae; the spaces between the trabeculae are filled with red marrow.

5. Ligaments attach bone to bone, and tendons attach muscle to bone.

6. The skeletal system contains three types of cartilage: fibrocartilage, hyaline cartilage, and elastic cartilage.

7. Ossification is the process of developing bone from cartilage, which begins in the embryo.

8. Bones lengthen through adolescence, and undergo replacement, repair, and remodeling throughout adulthood.

9. Bone cells include osteocytes that maintain bone tissue, osteoblasts that build bone tissue, and osteoclasts that dissolve bone tissue.

10. Homeostasis of bone tissue requires a balance in the activity of osteoblasts and osteoclasts.

11. Osteoblast and osteoclast activity is regulated by parathyroid hormone and calcitonin.

Sections 5.4, 5.5, 5.6

12. Bones are classified by shape as long, short, flat, and irregular.

13. Bones consist of both compact and spongy bone tissue arranged to meet the functional needs of the bone.

14. The skeleton is organized into the axial skeleton and the appendicular skeleton.

15. The axial skeleton consists of the skull, vertebral column, ribs, and sternum.

16. The appendicular skeleton consists of the pectoral girdle, pelvic girdle, and limbs.

17. Joints are points of contact between bones and are called articulations.

18. Joints are classified in three broad categories based on mobility: fibrous joints are immovable, cartilaginous joints are slightly movable, and synovial joints are freely movable.

19. Synovial joints contain a joint cavity lined with synovial membrane and filled with synovial fluid.

20. Diseases and disorders of the skeletal system include sprains, bursitis, tendinitis, osteoarthritis, rheumatoid arthritis, and osteoporosis.

Exercises

Complete the exercises for each section after you have read and studied the section. If you cannot answer some questions, or if you answer them incorrectly, return to the chapter and review this information. You may find it helpful to work on only one section at a time. When you have completed all sections, take the Chapter Test as an indicator of your mastery of this topic.

5.1 The skeletal system consists of connective tissue

5.2 Bone development begins in the embryo

5.3 Mature bone undergoes remodeling and repair

Matching

Select from the following list the correct term for each definition below.

growth plate	**bone**	**osteons**	**central canal**
cartilage	**chondroblasts**	**ligaments**	

_____ 1. dense fibrous connective tissue structures that attach bone to bone

_____ 2. a strip of cartilage in the epiphysis of a long bone that allows for bone lengthening

_____ 3. the hard elements of the skeleton

_____ 4. cartilage-forming cells

_____ 5. a ringlike arrangement of bone tissue formed in compact bone

_____ 6. a passageway through the center of an osteon that carries blood vessels

_____ 7. a connective tissue that contributes to the function of the skeletal system by forming embryonic structures and aiding movement

Short Answer

8. What are the three components of the skeletal system?

9. List five important functions of bones and the skeletal system.

10. What are trabeculae and where are they found?

11. How does red marrow differ from yellow marrow in location, composition, and function?

12. How do osteocytes in compact bone receive nutrients once they become trapped in lacunae?

Fill-in-the-Blank

Referenced sections are in parentheses.

13. A mass of clotted blood produced when a bone fractures is a _____. (5.3)

14. A callus is a _____ bond between broken ends of bone, formed by _____ during bone repair. (5.3)

Completion

Complete the table by filling in the missing information about cartilage.

Type of cartilage	Primary fiber type	Characteristics	Location
15. a)	b)	c)	Intervertebral discs, menisci of the knee
16. a)	b)	Creates a smooth, low-friction surface	c)
17. a)	Elastin	b)	c)

Labeling

Identify each structure indicated in Figure 5.1.

Figure 5.1

18. _____	24. _____
19. _____	25. _____
20. _____	26. _____
21. _____	27. _____
22. _____	28. _____
23. _____	29. _____

Ordering

Number each of these events, placing them in the proper sequence with the first event numbered "1" and the last event numbered "8."

30. Embryonic bone development:

_____ a. osteoblasts develop into osteocytes

_____ b. bone models are constructed of hyaline cartilage

_____ c. osteoblasts secrete osteoid

_____ d. blood vessels enter the area, bringing osteoblasts

_____ e. chondroblasts die

_____ f. osteocytes maintain bone matrix

_____ g. hydroxyapatite develops in osteoid matrix

_____ h. cartilage matrix breaks down

31. Bone repair:

_____ a. osteoblasts arrive in the area

_____ b. a hematoma develops

_____ c. fibroblasts become chondroblasts

_____ d. blood vessels bleed into the broken or fractured area

_____ e. osteoclasts arrive and destroy bone fragments

_____ f. new bone tissue is produced

_____ g. a callus forms

_____ h. fibroblasts migrate to the injured area

Word Choice

Circle the correct term to complete each sentence.

32. Spongy bone is (more/less) dense than compact bone.

33. (Osteoblasts/Osteocytes/Osteoclasts) break down bone tissue.

34. During bone remodeling, new bone is produced in areas of (low/high) compressive stress.

35. Mature compact bone is (one-third/two-thirds) osteoid and (one-third/two-thirds) hydroxyapatite.

36. Once cartilage in the growth plate has been replaced by bone tissue, the bone can continue to grow only in (length/width).

37. Falling levels of calcium in the blood (increase/decrease) the activity of parathyroid hormone.

38. Rising levels of calcium in the blood (increase/decrease) the activity of calcitonin.

39. Calcitonin (increases/decreases) the activity of osteoblasts.

40. Increased osteoclast activity (increases/decreases) the blood level of calcium.

5.4   The skeleton protects, supports, and permits movement

5.5   Joints form connections between bones

5.6   Diseases and disorders of the skeletal system

Labeling

Identify each of the bones indicated in Figure 5.2.

Figure 5.2

1. _____   9. _____   17. _____

2. _____   10. _____   18. _____

3. _____   11. _____   19. _____

4. _____   12. _____   20. _____

5. _____   13. _____   21. _____

6. _____   14. _____   22. _____

7. _____   15. _____   23. _____

8. _____   16. _____

Completion

Name each injury or disorder of the skeletal system described in the table.

Injury or disorder of the skeletal system	Description
24.	inflammation of the sinuses
25.	excessive loss of bone mass
26.	swelling and inflammation of tendons, causing pressure on the nerves that supply the hand
27.	joint inflammation caused by a malfunction of the immune system
28.	stretched or torn ligaments accompanied by bruising, swelling, and pain
29.	joint inflammation resulting from degenerative wear and tear
30.	inflammation of the bursae following injury

Crossword Puzzle

Across

2. The _____ pubis is a cartilaginous joint between the coxal bones
4. A joint _____ is found only in synovial joints
5. A band of connective tissue that connects bone to bone
7. A point of contact between two vertebrae
9. Cartilage pads that cushion the area between two vertebrae
10. This type of joint is the most freely movable
12. Sacs of fluid that provide cushioning within a joint
13. The _____ girdle supports the upper limbs

Down

1. This region of the vertebral column contains seven vertebrae
2. The flexibility of the human skeleton results in a decrease in _____
3. Immovable joints
6. A band of connective tissue that connects muscle to bone
8. This part of the skeleton includes the skull and vertebral column
10. This type of bone is as wide as it is long
11. This region of the vertebral column contains five vertebrae

Looking Back/Looking Forward

1. Along with disrupting growth, why is damage to the growth plate a serious injury? (Refer to section 4.3.)

2. Chapter 13 discusses the functions and regulation of hormones in the body, including parathyroid hormone (PTH). In addition to stimulating osteoclasts, PTH plays a critical role in vitamin D activation, a vitamin required for the absorption of calcium by the digestive tract. Children with insufficient levels of vitamin D may develop rickets, a disorder in which the bones are too soft. How might low PTH levels lead to rickets?

Chapter Test

Multiple Choice

1. Which of the following is not a function of both bone and the skeleton?
   a. support
   b. movement
   c. protection
   d. mineral storage

2. Which of the following correctly represents the three types of connective tissue that make up the skeletal system?
   a. bones, matrix, adipose tissue
   b. ligaments, bones, cartilage
   c. collagen, ligaments, cartilage
   d. ligaments, cartilage, joints

3. Red bone marrow
   a. fills the central cavity of the shaft of a long bone.
   b. contains primarily fat cells.
   c. is located in the spaces between the trabeculae of spongy bone.
   d. is transitional tissue found in the growth plate.

4. Canaliculi
   a. are transitional cells formed during ossification.
   b. are thin rods of bone found only in spongy bone tissue.
   c. run through compact bone at the center of each osteon.
   d. provide contact channels for osteocytes.

5. Spongy bone and compact bone tissue both consist of a hard mineralized matrix containing calcium and phosphate.
   a. true
   b. false

6. A structure composed of dense fibrous connective tissue that attaches bone to bone is a
   a. ligament.
   b. tendon.
   c. osteon.
   d. trabeculae.

7. Hyaline cartilage is found in
   a. intervertebral discs.
   b. the ends of long bones.
   c. the ear and epiglottis.
   d. tendons and ligaments.

8. Which of the following occurs first in the process of ossification?
   a. Blood vessels develop in the cartilage model.
   b. Osteoblasts begin building bone tissue.
   c. Chondroblasts die.
   d. Growth plates allow lengthening.

9. A mature bone that had a lower-than-normal percentage of hydroxyapatite would be
   a. too flexible.
   b. too brittle.
   c. too long.
   d. too short.

10. Replacement of cartilage in the growth plate with bone tissue
    a. only occurs when a bone has completed all growth and remodeling.
    b. is triggered by rising levels of PTH.
    c. indicates an imbalance in calcitonin and PTH.
    d. occurs when levels of growth hormone are too high.
    e. prevents further lengthening of the bone.

11. What is the role of an osteocyte in the body?
    a. to dissolve bone tissue
    b. to balance the activity of osteoblasts and osteocytes
    c. to maintain bone tissue
    d. to produce calcitonin and PTH

12. The bones of trained athletes may be thicker and heavier than the bones of nonathletes because
    a. athletes generally have better nutrition.
    b. athletic activity stimulates higher hormone levels.
    c. bone mass increases when bone is stressed.
    d. athletic activity decreases the activity of osteocytes.

13. Parathyroid hormone levels increase when
    a. osteoclast activity is too high.
    b. blood levels of calcium are too low.
    c. blood levels of calcium are too high.
    d. new bone tissue is needed.

14. Adequate calcium intake is important in the prevention of osteoporosis. Which of the following might occur if calcium intake was too low?
    a. Low blood calcium levels would stimulate the activity of osteoblasts.
    b. Low blood calcium levels would stimulate the activity of osteoclasts.
    c. Inadequate calcium would be available for deposit into bone tissue.
    d. Osteocytes would die.
    e. b and c.

15. A callus is produced
    a. when osteoclasts break down bone tissue during bone remodeling.
    b. during bone repair as a bridge between two broken edges of a bone.
    c. by red blood cells released into a damaged area of bone.
    d. by chondroblasts during bone lengthening.

16. Fused vertebrae are found naturally in the _____ region of the vertebral column.
    a. lumbar
    b. coccygeal
    c. thoracic
    d. cervical

17. A bone with only a small amount of spongy bone sandwiched between two layers of compact bone would be classified as
    a. short.
    b. flat.
    c. long.
    d. irregular.

18. Which of the following is not part of the pelvic girdle?
    a. coxal bone
    b. clavicle
    c. sacrum
    d. coccyx

19. Which of the following bones are not correctly identified?
    a. phalanges - fingers and toes
    b. ulna - forearm
    c. femur - upper leg
    d. tarsals - wrist

20. The disadvantage of highly flexible joints is
    a. inadequate blood supply.
    b. poor control of movement.
    c. poor stability.
    d. limited movement.

21. Which of the following correctly sequences the joints in order of increasing mobility?
    a. synovial, fibrous, cartilaginous
    b. fibrous, synovial, cartilaginous
    c. cartilaginous, fibrous, synovial
    d. fibrous, cartilaginous, synovial

22. If a knee joint lacked menisci it would be less able to
    a. attach to the muscles responsible for movement.
    b. secrete synovial fluid.
    c. repair damaged cartilage.
    d. reduce friction during movement.

23. A movement that increases the angle of a joint is called
    a. supination.
    b. flexion.
    c. extension.
    d. circumduction.

24. Inflammation of the joints due to an autoimmune response is called
    a. osteoarthritis.
    b. immunoarthritis.
    c. rheumatoid arthritis.
    d. inflammatory arthritis.

25. Which of the following is *not* a possible cause of bursitis?
    a. excessive osteoclast activity
    b. torn ligaments
    c. bacterial infection
    d. inflammation

Self Test

Each of the Key Concepts listed at the beginning of this chapter has been rewritten as a question below. After successfully completing the study guide exercises and the Chapter Test, you should be able to answer each of these questions. Refer to the Key Concepts list at the beginning of this chapter to check your answers.

1. What three types of connective tissue are found in the skeletal system?

2. What are three functions of the skeleton and two additional functions of bone?

3. How would you describe the structure of compact bone?

4. How would you describe the structure of spongy bone?

5. What is the function of a ligament? What is the function of a tendon?

6. What three types of cartilage are found in the skeletal system?

7. What is ossification? When does it begin?

8. Which bone activities occur in both adolescence and in adulthood?

9. What are the names of three types of bone cells, and what is the function of each?

10. What is required for homeostasis of bone tissue?

11. What two hormones regulate the activity of osteoblasts and osteoclasts?

12. What are the four categories of shape used to classify bones?

13. What type of bone tissue is found in each shape of bone?

14. What are the two divisions of the skeleton?

15. What is included in the axial skeleton?

16. What is included in the appendicular skeleton?

17. What is a joint? What is another term for a joint?

18. What are the names of the three categories of joints, and what is the mobility of each?

19. How would you describe the structure of a synovial joint?

20. What are the five disorders of the skeletal system discussed in this chapter?

6

The Muscular System

Chapter Summary and Key Concepts

After reading and studying this chapter you should know the following:

Sections 6.1, 6.2

1. Muscles produce movement, generate tension, and generate heat.

2. The three types of muscle are skeletal muscle, cardiac muscle, and smooth muscle.

3. Muscle cells are excitable and act by contracting and then relaxing.

4. Muscles most frequently attach to bone via tendons, but they may also attach to other muscles or to the skin.

5. Muscles have one attachment point called the origin and a second attachment point called the insertion. When muscles contract, the insertion is pulled toward the origin.

6. In skeletal muscle, individual muscle cells are called fibers, and they are arranged in bundles called fascicles. A whole muscle contains many fascicles bound together with connective tissue. The connective tissue layers of a muscle extend beyond the muscle as a tendon.

7. Muscle fibers contain myofibrils, which in turn contain the contractile filaments actin and myosin.

8. The contractile unit of a muscle is the sarcomere.

9. Skeletal muscle cells are stimulated by motor neurons that release the neurotransmitter acetylcholine.

10. During contraction, acetylcholine stimulates the release of calcium ions into the cytoplasm. The presence of calcium triggers the events leading to contraction.

11. Contraction results when actin filaments are pulled across myosin filaments, causing the sarcomere to shorten.

12. Muscle contraction and relaxation require ATP.

Section 6.3

13. The force of muscle contraction is determined by the number of muscle cells in a motor unit, the number of motor units involved, and the frequency of nerve stimulation.

14. A twitch is a complete cycle of contraction and relaxation. A latent period precedes a twitch.

15. Recruitment is an increase in muscle force due to an increasing number of motor units contracting.

16. Summation is an increase in muscle force due to an increase in the frequency of nerve stimulation.

17. The internal structure and chemistry of a muscle fiber will determine whether it is a slow-twitch fiber or a fast-twitch fiber. Muscle strength and endurance are determined by the number of slow-twitch and fast-twitch fibers present.

Sections 6.4, 6.5

18. Cardiac muscle cells have one nucleus, intercalated discs, and striations; are involuntary; and are found only in the heart.

19. Smooth muscle cells have one nucleus, gap junctions, and no striations; are involuntary; and are found around hollow organs or tubes.

20. Disorders of the muscular system include muscular dystrophy, tetanus, muscle cramps, pulled muscles, and fasciitis.

Exercises

Complete the exercises for each section after you have read and studied the section. If you cannot answer some questions, or if you answer them incorrectly, return to the chapter and review this information. You may find it helpful to work on only one section at a time. When you have completed all sections, take the Chapter Test as an indicator of your mastery of this topic.

6.1 Muscles produce movement or generate tension

6.2 Individual muscle cells contract and relax

Matching

Select from the following list the correct term for each definition below.

skeletal muscle	**neurotransmitter**	**origin**
motor neuron	**neuromuscular junction**	**insertion**
oxygen debt	**sarcoplasmic reticulum**	**myofibrils**
sarcomere	**sliding filament mechanism**	**actin**
fatigue	**myosin**	

_____ 1. the contractile unit of a muscle cell

_____ 2. a process resulting in shortening of a sarcomere

_____ 3. a decline in muscle performance during sustained exercise

_____ 4. muscles that attach to bone

_____ 5. a chemical messenger molecule that stimulates muscle cells to contract

_____ 6. a specialized smooth endoplasmic reticulum that stores calcium ions in muscle cells

_____ 7. point of attachment of a skeletal muscle to a structure that remains stationary

_____ 8. a condition that occurs when muscles use more ATP than can be supplied by aerobic respiration

_____ 9. cylindrical structures within a muscle cell that are packed with actin and myosin

_____ 10. the protein found in thick filaments

_____ 11. point of attachment of a skeletal muscle to a structure that moves

_____ 12. the junction between a motor neuron and a skeletal muscle cell

_____ 13. the protein found in thin filaments

_____ 14. releases acetylcholine

Word Choice

Circle the correct term to complete each sentence.

15. When muscle cells contract they (shorten/lengthen), and when they relax they (shorten/lengthen).

16. Synergistic muscles work to perform (the same/opposing) movements.

17. When a muscle contracts, the (origin/insertion) is pulled toward the (origin/insertion).

18. Using the midline of the body as a reference, the origin of a muscle is usually (closer to/farther from) the midline than the insertion.

19. Maintaining proper posture requires many muscles to (produce/resist) movement.

20. When muscle cells respond to chemical signals they are referred to as (electrical/excitable).

21. A whole muscle is a group of individual muscle cells with (the same/different) origin(s) and insertion(s).

22. The layers of fascia within a single muscle come together at the end of the muscle to form a (tendon/ligament).

23. A single skeletal muscle cell contains (one nucleus/multiple nuclei).

24. A single muscle cell contains (myofibrils packed with actin and myosin/actin and myosin packed with myofibrils).

Labeling/Coloring

Select different colors for each muscle listed below. Color the coding circles and the muscle(s) in Figure 6.1 (not all muscles will be colored). Then, label the muscles and write the function of each muscle in Figure 6.1 in the spaces provided. After completing this exercise, refer to Figure 6.2 in your text to check your work.

○ Gluteus maximus   ○ Pectoralis major

○ Sartorius   ○ Quadriceps group

○ Tibialis anterior   ○ Rectus abdominus

○ Latissimus dorsi   ○ Gastrocnemius

○ Hamstring group   ○ Biceps brachii

○ Deltoid   ○ Trapezium

Figure 6.1

Ordering

Number each of the following events of muscle contraction to indicate the sequence in which they occur.

_____ 25. the troponin-tropomyosin complex shifts position, exposing myosin-binding sites

_____ 26. an electrical impulse is carried deep into muscle cells by the T-tubules

_____ 27. acetycholine is released at the neuromuscular junction

_____ 28. the sliding filament mechanism results in shortening of a sarcomere

_____ 29. cross-bridges between myosin and actin

_____ 30. calcium ions bind to troponin

_____ 31. calcium ions are released from the sarcoplasmic reticulum

Completion

Answer the following questions about Figure 6.2. Use the alphabetized labels in the figure to indicate your answers.

_____ 32. represents myosin

_____ 33. associated with the troponin-tropomyosin complex

_____ 34. has heads that form cross-bridges

_____ 35. slides toward the center of the sarcomere during a contraction

_____ 36. has myosin-binding sites

_____ 37. represents the Z-line

Figure 6.2

Short Answer

38. Explain two ways that ATP is required for muscle relaxation.

39. List the four sources of ATP available to a muscle cell in the order in which the ATP will be accessed.

6.3 The activity of muscles can vary

Define

Write the correct term(s) for each definition:

_____ 1. a neuron that affects movement

_____ 2. a motor neuron and all the muscle cells it controls

_____ 3. the mechanical force a muscle generates during contraction

_____ 4. a principle stating that muscle cells will always respond fully to a stimulus from a motor neuron

_____ 5. a complete cycle of contraction and relaxation

_____ 6. increasing muscle force by increasing the number of motor units active at one time

_____ 7. a laboratory recording of muscle activity

_____ 8. increasing muscle force by increasing the rate of stimulation of motor units

Short Answer

9. Characterize isotonic and isometric contractions in terms of force generated, length of the muscle, and movement.

10. What three factors determine the amount of tension generated by a muscle?

11. Describe the relationship between motor unit size, strength, and degree of control during muscle contraction.

12. Why does summation increase muscle force during a contraction?

13. What causes tetanus?

Fill-in-the-Blank

14. Muscle _____ is a minimal level of contraction present in most muscles when some motor units are contracting and others are not.

15. Muscle fibers that rely on aerobic metabolism for energy are referred to as _____ fibers.

16. Muscle fibers that produce a powerful, but short, contraction are referred to as
_____ fibers.

17. In exercise, _____ training increases the body's efficiency in delivering
oxygen to muscles.

18. The two primary types of exercise training are _____ and
_____.

Completion

Fill in the missing information for each section of the table.

19.

Stage of twitch	Description	Events
Latent period	a)	b)
c)	d)	Actin filaments are pulled toward the center of the sarcomere
e)	The muscle returns to its original length	f)

Critical Thinking

20. Why does strength training have a greater effect on fast-twitch fibers than slow-twitch fibers?

21. Why do most marathon runners have a lean build, while weight lifters exhibit greater muscle mass?

22. A friend has just joined the gym and excitedly explains to you that he is taking amino acid supplements to help him make more muscle cells, which he hopes will result in an increase in muscle mass. How would you respond?

6.4 Cardiac and smooth muscles have special features

6.5 Diseases and disorders of the muscular system

Matching

List each muscle type beside all applicable characteristics. Muscle types may be used more than once, and several muscle types may correspond with a characteristic.

a. skeletal muscle   **b. cardiac muscle**   **c. smooth muscle**

_____ 1. voluntary

_____ 2. involuntary

_____ 3. has intercalated discs

_____ 4. striated

_____ 5. non-striated

_____ 6. joined by gap junctions

_____ 7. attached to bones

_____ 8. can only contract when stimulated by a nerve cell

Short Answer

For each muscular disorder listed below, indicate the cause and the symptoms.

9. Muscular Dystrophy:

    a. Cause:

    b. Symptoms:

10. Tetanus:

    a. Cause:

    b. Symptoms:

11. Muscle Cramps:

    a. Cause:

    b. Symptoms:

12. Pulled Muscle:

    a. Cause:

    b. Symptoms:

13. Fasciitis:

    a. Cause:

    b. Symptoms:

Looking Back/Looking Forward

1. Muscle contraction is initiated when the Ca^{++} diffuses from the sarcoplasmic reticulum of a muscle cell into the cytoplasm. When the muscle is at rest, Ca^{++} move back into the sarcoplasmic reticulum where they are stored until the muscle is again stimulated by a motor neuron. What membrane transport process will be responsible for the movement of Ca^{++} back in to the sarcoplasmic reticulum?

2. Chapter 9 will discuss the Immune System and body defenses. Myasthenia gravis is an autoimmune disorder in which the body mistakenly attacks the acetylcholine receptors on skeletal muscle cells, resulting in a decrease in the number of receptors. What effect will this have on the muscle cells involved in the disease? Explain your answer.

Chapter Test

Multiple Choice

1. Muscle movements over which we have no conscious control are
   a. voluntary.
   b. involuntary.
   c. subconscious.
   d. reflexive.

2. The biceps brachii bends the forearm at the elbow and the triceps brachii extends the forearm at the elbow. These muscles are
   a. antagonistic.
   b. synergistic.
   c. complementary.
   d. cooperative.

3. When a skeletal muscle contracts, the _____ is pulled toward the _____.
   a. origin, insertion
   b. origin, muscle
   c. insertion, origin
   d. insertion, muscle

4. Individual muscle cells within a whole muscle are arranged in bundles called
   a. fascia.
   b. fascicles.
   c. fibers.
   d. fibrils.

5. When a muscle contracts, shortening of the muscle occurs because
   a. myofibrils slide over each other.
   b. T tubules shorten.
   c. sarcomeres shorten.
   d. muscle fascia constricts.

6. In the sliding-filament mechanism of muscle contraction,
   _____ slides over _____.
   a. Z-lines, actin
   b. actin, myosin
   c. myosin, actin
   d. Z-lines and actin, myosin

7. Which of the following would interfere most with a muscle's ability to contract?
   a. absence of acetylcholine-binding sites on the muscle membrane
   b. a reduced number of actin and myosin filaments
   c. an increased level of intracellular calcium
   d. a reduced number of T tubules

8. In a muscle cell, _____ contain _____.
   a. myofibrils, actin and myosin filaments
   b. actin and myosin filaments, myofibrils
   c. sarcomeres, myofibrils
   d. myofibrils, sarcomeres
   e. a and d

9. Which of the following best describes the action of sarcomeres within a muscle cell during a contraction?
   a. Each sarcomere shortens a little.
   b. Only the sarcomeres on the ends of the muscle shorten in a weak contraction.
   c. Sarcomeres in the center of the muscle shorten more than sarcomeres on the ends of the muscle.
   d. Sarcomeres shorten to different degrees, depending on the amount of actin and myosin they contain.

10. The force of a muscle contraction can be increased by
    a. increasing the frequency of nerve stimulation.
    b. decreasing the number of motor units responding.
    c. increasing the amount of collagen in the tendons.
    d. decreasing the amount of Ca^{++} released from the sarcoplasmic reticulum.

11. Place the following events of muscle contraction in the proper sequence.
    I. Calcium is released from the sarcoplasmic reticulum.
    II. Cross-bridges form.
    III. Actin slides across myosin.
    IV. Calcium binds to troponin, causing the troponin-tropomyosin complex to shift its position.

    a. I, II, IV, III

    b. I, IV, II, III

    c. II, I, IV, III

    d. III, I, IV, II

12. In the formation of a cross-bridge, _____ binds to

    _____ .

    a. myosin, actin

    b. actin, troponin

    c. myosin, tropomyosin

    d. tropomyosin, actin

13. When a muscle relaxes, ATP is required because

    a. the actin filaments must be pushed back to their original position.

    b. the myosin heads can only detach from the actin when a molecule of ATP binds to the head.

    c. energy is needed to pull the sarcomeres apart.

    d. the connective tissue elements of the muscle require energy to resume their original position.

14. Which of the following answers represents the correct order in which a muscle cell will access energy sources?

    a. creatine phosphate, stored ATP, glycogen, aerobic metabolism

    b. aerobic metabolism, creatine phosphate, glycogen, stored ATP

    c. stored ATP, glycogen, creatine phosphate, aerobic metabolism

    d. stored ATP, creatine phosphate, glycogen, aerobic metabolism

15. A muscle contraction that does not result in movement is called

    a. isometric.

    b. isotonic.

    c. tetanic.

    d. metabolic.

16. The physiological cause of muscle fatigue is

    a. inadequate sarcomeres to meet muscle demands.

    b. inadequate ATP to meet metabolic demands.

    c. inadequate numbers of slow-twitch fibers.

    d. constriction of blood vessels as lactic acid accumulates.

17. A motor unit consists of

    a. one motor neuron and all the muscle cells it stimulates.

    b. one muscle and all the motor neurons that stimulate that muscle.

    c. one motor neuron and one muscle cell.

    d. all the motor neurons and all the muscle cells of one whole muscle.

18. The all-or-none principle states that
    a. when a muscle is stimulated, every motor unit will respond.
    b. when a muscle cell is stimulated, it responds with a complete cycle of contraction and relaxation.
    c. when a muscle cell is stimulated, the sarcomeres that respond will shorten.
    d. a small stimulus can produce a small contraction.

19. Increasing muscle force by increasing the number of motor units responding is called
    a. summation.
    b. recruitment.
    c. tetany.
    d. magnification.

20. Tetanic contractions result from
    a. more than one twitch occurring at the same time.
    b. an action potential increasing the size of a muscle's motor units.
    c. myosin and actin being unable to bind to each other.
    d. frequent stimulation preventing a muscle from relaxing.

21. Which of the following is the same in both slow-twitch and fast-twitch muscle fibers?
    a. the all-or-none principle
    b. the role of aerobic metabolism
    c. the number of mitochondria
    d. the arrangement of actin and myosin in a sarcomere
    e. a and d

22. Strength training involves
    a. improving muscle endurance.
    b. providing resistance that requires muscles to work harder.
    c. increasing the body's ability to supply oxygen to the muscles.
    d. increasing the number of muscle cells.

23. Cardiac muscle and smooth muscle are similar in all of the following ways *except*
    a. both have muscle cells joined by gap junctions.
    b. the contraction of both can be modified by nerve stimulation.
    c. both are involuntary muscles.
    d. both fatigue easily.

24. Which of the following disorders of the muscular system is incorrectly described?
    a. tetanus—a bacterial disease that causes forceful muscle contractions
    b. muscular dystrophy—a viral disease that causes muscle wasting
    c. pulled muscle—muscle fibers tear apart due to excessive stretching of the muscle
    d. fasciitis—inflammation of the connective tissue sheath around a muscle

25. Internal bleeding, swelling, and pain occur with
    a. tetanus.
    b. muscle cramps.
    c. pulled muscles.
    d. fasciitis.

Self Test

Each of the Key Concepts listed at the beginning of this chapter has been rewritten as a question below. After successfully completing the study guide exercises and the Chapter Test, you should be able to answer each of these questions. Refer to the Key Concepts list at the beginning of this chapter to check your answers.

1. What are three things muscles can do?

2. What are the three types of muscle tissue?

3. How do muscle cells respond when they are stimulated?

4. What structures can be attachment points for muscles?

5. What are the origin and insertion? How do their positions change when a muscle contracts?

6. How would you describe the anatomy of a skeletal muscle?

7. Where are myofibrils located? What do myofibrils contain?

8. What is the contractile unit of a muscle cell?

9. What is the role of a motor neuron in stimulating skeletal muscle cells?

10. What is the effect of acetylcholine on a muscle cell?

11. What causes a sarcomere to shorten? What happens to a muscle cell when the sarcomere shortens?

12. Is ATP required for muscle contraction and relaxation?

13. What affects the force of a muscle contraction?

14. What is a twitch? What precedes a twitch?

15. What is recruitment?

16. What is summation?

17. What characterizes a muscle fiber as slow-twitch or fast-twitch? What qualities of the muscle are affected by the number of slow-twitch and fast-twitch fibers?

18. What are the characteristics of cardiac muscle?

19. What are the characteristics of smooth muscle?

20. What are five disorders of the muscular system described in this chapter?

7

Blood

Chapter Summary and Key Concepts

After reading and studying this chapter you should know the following:

Section 7.1

1. The functions of blood include transport, regulation, and defense.

2. Blood is composed of plasma and formed elements.

3. Plasma is mostly water and also contains ions, proteins, hormones, gases, nutrients, and wastes.

4. Important plasma proteins include albumins that maintain water balance, globulins that transport substances and function as antibodies, and clotting proteins.

5. Formed elements include red blood cells, called erythrocytes; white blood cells, called leukocytes; and platelets.

6. All formed elements are produced by stem cells in the bone marrow.

7. Mature red blood cells have no nucleus, and they contain an oxygen-binding protein called hemoglobin that transports oxygen from the lungs to the body, and some carbon dioxide from the body to the lungs.

8. The liver and spleen destroy old and damaged red blood cells, recycling the iron atoms and converting heme groups to bilirubin.

9. Red blood cell production is controlled by the hormone erythropoietin.

10. White blood cells defend the body. They are classified as granular or agranular depending on their staining properties.

11. Granular leukocytes include neutrophils, eosinophils, and basophils. Agranular leukocytes include monocytes and lymphocytes.

12. Platelets are essential to the process of blood clotting. Their production is regulated by the hormone thrombopoietin.

Section 7.2

13. Hemostasis is the process of stopping blood loss, which occurs through vascular spasm; formation of a platelet plug; and coagulation.

14. Formation of a blood clot requires clotting factors, including prothrombin activator, thrombin, and fibrinogen.

Section 7.3

15. Blood type determines blood compatibility for transfusions.

16. Blood type is determined by cell surface proteins, called antigens, on the red blood cells.

17. Two common blood typing systems are the ABO system and the Rh system.

18. Transfusion reactions occur when foreign antigens are attacked by antibodies.

Section 7.4

19. Blood disorders may involve problems with any of the components of the blood.

20. Significant blood disorders include anemia, leukemia, multiple myeloma, mononucleosis, blood poisoning, and thrombocytopenia.

Exercises

Complete the exercises for each section after you have read and studied the section. If you cannot answer some questions, or if you answer them incorrectly, return to the chapter and review this information. You may find it helpful to work on only one section at a time. When you have completed all sections, take the Chapter Test as an indicator of your mastery of this topic.

7.1 The components and functions of blood

Define

Write the term(s) for each definition below:

_____ 1. a specialized connective tissue consisting of cells and cell fragments suspended in a watery solution

_____ 2. the liquid portion of blood

_____ 3. the largest group of solutes in plasma

_____ 4. plasma proteins that help maintain water balance in the body

_____ 5. plasma proteins that transport substances in the blood

_____ 6. red blood cells

_____ 7. white blood cells

_____ 8. an oxygen-binding protein found in red blood cells

_____ 9. hemoglobin with four oxygen molecules attached

_____ 10. the percentage of the blood consisting of red blood cells

_____ 11. cells in the red bone marrow that divide to produce all blood cells and platelets

_____ 12. large white blood cells that engulf and destroy debris and invaders by phagocytosis

_____ 13. the hormone that stimulates the production of red blood cells

_____ 14. artificially increasing the production of red blood cells

Short Answer

15. What are the three components of the circulatory system?

16. What three functions are served by the blood?

17. What is the average volume of blood in an adult male? In an adult female?

18. List the liquid components of the blood.

19. Describe the structure of a hemoglobin molecule.

20. What color is venous blood? Why?

Fill-in-the-Blank

Referenced sections are indicated in parentheses.

21. Mature red blood cells lack a _____ and most _____. (7.1)

22. Worn-out red blood cells are destroyed by the _____ and the _____. (7.1)

23. Immature cells that give rise to erythrocytes are called _____. (7.1)

24. Platelet production is regulated by the hormone _____. (7.1)

25. The category of agranulocytes that defend the body against large parasites is _____.

26. The inflammatory response is initiated by release of _____ from _____. (7.1)

27. Large white blood cells that differentiate into macrophages are _____. (7.1)

28. _____ are agranulocytes important to the immune system and are classified as B cells or T cells. (7.1)

29. When red blood cells are destroyed in the liver, the heme groups are converted to _____. (7.1)

30. Platelets are derived from large cells called _____. (7.1)

Word Choice

Circle the accurate term or phrase to complete each sentence.

31. Mature red blood cells have (no/multiple) nuclei.

32. The bond hemoglobin forms with oxygen is (temporary/permanent).

33. Hemoglobin binds O_2 best when the pH is (neutral/acidic/base).

34. Hemoglobin releases O_2 into the body tissues when the pH of the body tissues is (lower than/higher than) the pH of the blood.

35. Carbon dioxide (attaches to/detaches from) hemoglobin in the lungs.

36. CO_2 and O_2 bind to hemoglobin at (the same site/different sites).

37. Most white blood cells have a (shorter/longer) life span than a red blood cell.

38. Platelet production is regulated by the hormone (thrombopoietin/erythropoietin).

39. (White blood cells/Red blood cells) are part of the body's defense system.

Completion

Fill in the requested information for each type of white blood cell listed in the table below.

Mature White Blood Cell	What type of stem cell gives rise to this cell?	Is this cell granular or agranular?	What function will this cell serve?
40. Neutrophil			
41. Eosinophil			
42. Basophil			
43. Monocyte			
44. Lymphocyte			

Short Answer

45. Explain how structure and function are related in red blood cells.

46. Compare and contrast red blood cells and white blood cells by listing two ways that they are similar and two ways that they are different.

Critical Thinking

47. Why is the number of red blood cells monitored through oxygen levels in the blood rather than according to the number of red blood cells?

48. If an individual with low oxygen levels produced excessive amounts of erythropoietin, what effect would this have on the number of red blood cells produced?

7.2   Hemostasis: Stopping blood loss

True/False

Mark each statement as true or false. If you mark a statement as false, replace the underlined term or phrase with the term or phrase needed to make the statement true.

_____   1. The <u>first</u> stage of hemostasis is vascular spasm.

_____   2. Clotting factors are important in the formation of a <u>platelet plug</u>.

_____   3. Thrombin converts <u>fibrin to fibrinogen</u>.

_____   4. The third stage of hemostasis is <u>prothrombin activation</u>.

_____   5. A clot is formed when fibrin interacts with <u>platelets</u> and red blood cells.

7.3   Human blood types

Completion

Complete the table below with the information requested for each blood type.

Blood Type	Type of antigen on cell surface?	Plasma antibodies carried in the blood?	Can receive transfusions from what blood types?
1. Type A			
2. Type B			
3. Type AB			
4. Type O			

Fill-in-the-Blank

Referenced sections are indicated in parentheses.

5. Cells have surface _____ that are recognized by the immune system as "self." (7.3)

6. Foreign cells carry surface proteins called _____ that are recognized by the immune system as "nonself." (7.3)

7. When the immune system is exposed to foreign cells it will mount a defense that includes the production of _____ by _____. (7.3)

8. Antibodies float freely in the _____ and _____. (7.3)

9. When antibodies encounter a foreign antigen they will bind to it, forming a(n) _____. (7.3)

10. Throughout life, individuals carry circulating _____ against any antigens different from their own. (7.3)

11. When red blood cells with foreign antigens are attacked by _____, they clump together or _____. (7.3)

12. Clumping together of red blood cells (RBCs) may lead to _____ _____ or _____. (7.3)

13. An adverse effect of a blood transfusion is called a _____. (7.3)

Short Answer

14. Two individuals both have blood type A; however, one is type A+ and the other is type A–. What is the difference in their red blood cells?

15. Why can an individual with Rh– blood receive a one-time transfusion of Rh+ blood without complications, while a second transfusion of Rh+ blood may cause a severe reaction?

16. (a) What is hemolytic disease of the newborn? (b) What parental blood types would put a newborn at risk for this disease? (c) Why does this disease seldom affect a first child? (d) How can it be prevented in subsequent children?

17. Describe the process of cross-matching. What results indicate that two blood samples are a good match?

7.4   Blood disorders

Completion

For each disorder listed in the table below, describe the defining characteristic.

Disorder	Description
1. Mononucleosis	
2. Anemia	
3. Multiple myeloma	
4. Leukemia	
5. Blood Poisoning	
6. Thrombocytopenia	

Looking Back/Looking Forward

1. Oxygen and carbon dioxide diffuse easily in and out of red blood cells. How would it affect the body if these gases were moved by active transport rather than diffusion?

2. Chapter 18 discusses cell division and cancer. Cancer results from genetic damage that leads to uncontrolled cell division. Can mature red blood cells become cancerous? Explain your answer.

Chapter Test

Multiple Choice

1. The circulatory system consists of
   a.   red blood cells, white blood cells, and platelets.
   b.   plasma and formed elements.
   c.   blood vessels and lymphatic vessels.
   d.   heart, blood vessels, and blood.

2. Which of the following is *not* a function of the blood?
    a. protection against infection and illness
    b. regulation of body temperature and fluid balance
    c. secretion of erythropoietin
    d. transportation

3. Which of the following is *not* a component of plasma?
    a. platelets
    b. water
    c. dissolved protein
    d. ions

4. Which of the following plasma proteins is incorrectly matched with its function?
    a. albumin—helps to maintain proper water balance in the body
    b. beta globulin—creates lipoproteins
    c. gamma globulin—binds to fat molecules
    d. clotting protein—participates in blood clotting

5. A recent blood test indicates you have a high level of HDLs and a low level of LDLs. Your doctor tells you that you should be
    a. pleased—this indicates a reduced risk of cardiovascular disease.
    b. concerned—this indicates an increased risk of cardiovascular disease.
    c. cautious—LDL and HDL levels should be similar for optimal health.
    d. unconcerned—HDL and LDL levels have no impact on cardiovascular health.

6. Your friend has just returned from the doctor and tells you her hematocrit was 29%. You respond
    a. she is well within normal limits for an adult female.
    b. she should be concerned—29% is too low.
    c. her level of white blood cells may be too high.
    d. this value is insignificant; it is only the total blood volume that is important.

7. What conditions in the lungs contribute to the diffusion of oxygen into the blood?
    a. acidic pH and low O_2 concentration
    b. specialized transport proteins in the plasma membranes of lung cells
    c. high O_2 concentration and neutral pH
    d. high CO_2 concentration

8. Red blood cells are accurately described as
    a. large and oblong cells.
    b. rigid and inflexible cells.
    c. small and flat cells.
    d. multinucleate cells.

9. Stem cells in the bone marrow produce
   a. red blood cells only.
   b. white blood cells only.
   c. red and white blood cells.
   d. red cells, white cells, and platelets.

10. Red blood cells have a short life span because
    a. they are easily damaged as they pass through the blood vessels.
    b. they are often destroyed by immune system cells.
    c. they lack a nucleus.
    d. their cell membranes are impermeable to required nutrients.

11. Your friend shows you a yellow bruise on her arm. You explain to her that the yellow color is due to the
    a. reduced oxygen-carrying capacity of damaged red blood cells.
    b. breakdown of heme as red blood cells are destroyed.
    c. increased oxygen-binding capacity of hemoglobin during healing.
    d. high concentration of carbon dioxide at the site of the bruise.

12. Erythropoietin stimulates production of _____ in response to

    _____.
    a. white blood cells, infection
    b. red blood cells, low O_2 availability
    c. platelets, blood vessel damage
    d. fibrin, increased fibrinogen levels

13. Leukocytes differ from erythrocytes because only leukocytes
    a. arise from stem cells.
    b. have a nucleus.
    c. lack hemoglobin.
    d. b and c.

14. Eosinophils
    a. are the rarest type of white blood cell.
    b. defend the body against large parasites.
    c. are the first white blood cell to arrive at an injury site.
    d. are agranular leukocytes.

15. Which of the following conditions would increase the release of O_2 from red blood cells into the tissue fluids?
    a. neutral pH
    b. high concentration of O_2 in the tissue fluids
    c. low concentration of O_2 in the tissue fluids
    d. low concentration of CO_2 in the tissue fluids

16. In the process of hemostasis, when a blood vessel breaks,
    a. red blood cells differentiate into macrophages.
    b. eosinophils migrate to the site to seal the rupture.
    c. platelets swell, develop spiky extensions, and begin to clump together.
    d. the kidneys decrease production of erythropoietin.

17. In what order do the following substances participate in the formation of a blood clot?
    I.   fibrinogen
    II. thrombin
    III. prothrombin activator
    IV. fibrin
    V.   prothrombin
    a. IV, I, V, III, II
    b. I, IV, III, V, II
    c. II, V, III, IV, I
    d. III, V, II, I, IV

18. An individual who has blood type AB+ will have _____ antigens on the surface of their red blood cells.
    a. A and B
    b. only Rh
    c. A, B, and Rh
    d. no

19. Agglutination between the red blood cells of two blood samples indicates that
    a. the two samples are compatible.
    b. the cells of both samples have the same antigens.
    c. the two samples are not compatible.
    d. the antigens of one sample have attacked the antibodies of the other sample.

20. During a transfusion reaction,
    a. recipient antigens attack donor antibodies.
    b. recipient antigens fail to bind to donor antigens.
    c. recipient antibodies attack donor antigens.
    d. recipient antibodies engulf and destroy foreign red blood cells by phagocytosis.

21. In the ABO blood system, antibodies appear in an individual's blood
    a. only in response to foreign antigens.
    b. early in life, with or without exposure to foreign antigens.
    c. only during childhood.
    d. only in response to Rh+ blood.

22. Which of the following is *not* true of leukemia?
    a. it is a type of blood cancer
    b. too many white blood cells are produced
    c. the rate of red blood cell destruction is increased
    d. organ function is affected

23. Which type of anemia results from a deficiency in vitamin B_{12} absorption by the digestive tract?
    a. iron-deficiency anemia
    b. sickle-cell anemia
    c. aplastic anemia
    d. pernicious anemia

24. Mononucleosis
    a. is not contagious.
    b. is the result of a bacterial infection.
    c. has no known cure.
    d. symptoms include easy bruising and bone pain.

25. An individual who has blood type A can receive a transfusion of which blood type(s)?
    a. only type A
    b. type A and type O
    c. type A and type B
    d. type A and type AB
    e. type A, type AB, and type O

Self Test

Each of the Key Concepts listed at the beginning of this chapter has been rewritten as a question below. After successfully completing the study guide exercises and the Chapter Test, you should be able to answer each of these questions. Refer to the Key Concepts list on the first page of this chapter to check your answers.

1. What are the functions of the blood?

2. What are the two main components of blood?

3. What is plasma composed of?

4. What are three important categories of plasma proteins and their functions?

5. What are the formed elements of the blood?

6. Where are formed elements produced? What type of cell is involved?

7. What is found inside red blood cells? What is the function of hemoglobin?

8. Where and how are old red blood cells destroyed?

9. What hormone controls the production of red blood cells?

10. What is the function of white blood cells? How are they classified?

11. Which white blood cells are classified as granular leukocytes? Which white blood cells are classified as agranular leukocytes?

12. What process requires platelets? What hormone regulates platelet production?

13. What is hemostasis? What are the stages of hemostasis?

14. What substances are required for the formation of a blood clot?

15. What determines blood compatibility?

16. What determines an individual's blood type?

17. What are two commonly used blood typing systems?

18. What causes a transfusion reaction?

19. What parts of the blood may be involved in blood disorders?

20. What are the six blood disorders discussed in Chapter 7?

8

Heart and Blood Vessels

Chapter Summary and Key Concepts

After reading and studying this chapter, you should know the following:

Sections 8.1, 8.2

1. The pumping action of the heart provides the power to transport blood.

2. Arteries transport blood away from the heart; veins transport blood toward the heart.

3. Capillaries transport blood from arteries to veins, and they allow for exchange of fluid, nutrients, and wastes with the tissues.

4. The structure of the vessel wall is suited to its function.

5. Veins return blood to the heart with the aid of skeletal muscle activity, one-way valves, and pressure changes associated with breathing.

6. The lymphatic system helps to maintain blood volume by returning excess extracellular fluid to the blood.

7. The heart consists of four chambers—two upper atria that receive blood and two lower ventricles that eject blood from the heart.

8. Heart valves prevent blood from flowing backward.

9. The right side of the heart receives and then pumps deoxygenated blood. The left side of the heart receives and then pumps oxygenated blood.

10. The pulmonary circuit transports blood between the heart and the lungs for oxygenation.

11. The systemic circuit transports blood between the heart and the rest of the body to supply oxygen and nutrients to the tissues and remove wastes from the tissues.

12. The cardiac cycle includes periods of contraction called systole and periods of relaxation called diastole.

13. Electrical impulses cause contraction of heart muscle, and they are initiated and spread through the heart by the cardiac conduction system.

Sections 8.3, 8.4, 8.5, 8.6

14. Blood pressure is the force that blood exerts on the vessel walls.

15. Blood pressure is an important indicator of the overall health of the cardiovascular system.

16. Systolic pressure is produced when ventricles contract, and diastolic pressure is produced when ventricles relax.

17. A constant arterial blood pressure is crucial to homeostasis, and is maintained by regulating heart rate, force of contraction of the heart, and the diameter of arterioles.

18. Regulation of the cardiovascular system is accomplished by receptors in the arteries, regulatory centers in the brain, hormones, and the environment in the tissues.

19. Cardiovascular disorders include hypertension, aneurysms, atherosclerosis, angina pectoris, heart attack, congestive heart failure, embolism, and stroke.

20. There are many voluntary actions that can help to maintain cardiovascular health.

Exercises

Complete the exercises for each section after you have read and studied the section. If you cannot answer some questions, or if you answer them incorrectly, return to the chapter and review this information. You may find it helpful to work on only one section at a time. When you have completed all sections, take the Chapter Test as an indicator of your mastery of this topic.

8.1   Blood vessels transport blood

8.2   The heart pumps blood through the vessels

Define

Write the term(s) for each definition.

_____ 1. thick-walled vessels that transport blood at high pressure away from the heart

_____ 2. the smallest arteries of the body

_____ 3. band of smooth muscle that controls blood flow into the capillaries

_____ 4. the smallest vessels whose walls consist of a single layer of cells

_____ 5. vessels that transport blood from the body back to the heart

_____ 6. valves that prevent blood from flowing back into the atria when the ventricles contract

_____ 7. valves that prevent blood from flowing backward into the ventricles when the heart relaxes

_____ 8. blood vessels that supply the heart muscle

_____ 9. period of contraction in the cardiac muscle

_____ 10. period of relaxation of cardiac muscle

_____ 11. the complete sequence of contraction and relaxation in the atria and ventricles

_____ 12. starting location for the stimulus that initiates a heartbeat

_____ 13. the largest artery in the body

Fill-in-the-Blank

Referenced sections are indicated in parentheses.

14. The three major types of blood vessels in the body are _____, _____, and _____. (8.1)

15. Water and small solutes escape from a capillary through _____ in the epithelial cells and _____ between the cells. (8.1)

16. Ballooning of the arterial wall due to vessel damage is referred to as a(n) _____. (8.1)

17. Contraction of vascular smooth muscle is called _____. (8.1)

18. _____ of arterioles and precapillary sphincters increases blood flow to the capillaries. (8.1)

19. Fluid filtered from the beginning of a capillary is reabsorbed by _____ into the last part of the capillary. (8.1)

20. _____ _____ are extensive networks of capillaries found in all areas of the body. (8.1)

21. Filtration of fluid out of the capillaries is caused by _____ _____ generated by the heart. (8.1)

22. Blind-ended vessels that pick up excess extracellular fluid are _____ _____. (8.1)

23. The fluid carried by lymphatic vessels is called _____. (8.1)

24. Blood flows from capillaries into _____. (8.1)

25. Veins permanently swollen by pooled blood are called _____ veins. (8.1)

26. The _____ is a tough fibrous sac surrounding the heart. (8.2)

27. The _____ _____ contains fluid that provides a lubricant for contraction of cardiac muscle. (8.2)

28. An abnormality in the heart rate is called a(n) _____. (8.2)

29. Inflammation of the innermost layer of the heart is called _____. (8.2)

30. The _____ circuit sends deoxygenated blood to the lungs and returns oxygenated blood to the heart.

Labeling/Coloring

Select different colors for each structure listed below. Color the coding circles and the corresponding structures in Figure 8.1.

○ pulmonary trunk   ○ pulmonary semilunar valve

○ aortic semilunar valve   ○ right ventricle

○ left AV valve   ○ right AV valve

○ aorta   ○ septum

○ right atrium   ○ left ventricle

○ left atrium   ○ superior vena cava

Figure 8.1

Short Answer

31. List and describe the three layers of an artery.

32. List three functions of arterioles.

33. How does the structure of a capillary wall enable blood to exchange substances with tissue cells?

34. What three mechanisms help the veins return blood to the heart?

35. List the four structures of the cardiac conduction system.

36. List and describe the three layers of the heart wall.

37. What events in the heart are represented by each of the three formations in an ECG?

   a. P wave:

   _____

   b. QRS complex:

   _____

   c. T wave:

   _____

Word Choice

Circle the accurate term or phrase to complete each sentence.

38. During atrial systole
   a. both atria (contract/relax).
   b. the blood pressure in the atria (increases/decreases).
   c. the AV valves are (open/closed).
   d. the semilunar valves are (open/closed).

39. During ventricular systole
   a. the ventricles (contract/relax).
   b. the AV valves are (open/closed).
   c. the pulmonary and semilunar valves will (open/close).
   d. blood flows into the (pulmonary trunk and aorta/right and left atria).

40. During diastole
   a. the atria are (contracted/relaxed).
   b. the ventricles are (contracted/relaxed).
   c. the pulmonary and aortic semilunar valves will (open/close).
   d. the AV valves will (open/close).

Ordering

Order the following structures to indicate the correct sequence of blood flow, beginning and ending with the right atrium.

left atrioventricular valve

aorta

capillaries

arterioles

right ventricle

venules

left atrium

pulmonary veins

aortic semilunar valve

right atrioventricular valve

veins

arteries

lungs

pulmonary semilunar valve

pulmonary trunk

left ventricle

41. right atrium

→d. _____

→h. _____

→l. _____

→p. _____

→a. _____

→e. _____

→i. _____

→m. _____

→right atrium

→b. _____

→f. _____

→j. _____

→n. _____

→c. _____

→g. _____

→k. _____

→o. _____

Critical Thinking

42. If plasma proteins were able to diffuse out of a capillary, how would this affect the reabsorption of fluid back into the capillary before the capillary joins a venule?

43. How do the structural differences between arteries and veins reflect the different functions of these two vessels?

Labeling

Identify each blood vessel in Figure 8.2 on the next page.

8.3   Blood exerts pressure against vessel walls

8.4   How the cardiovascular system is regulated

8.5   Cardiovascular disorders: A major health issue

8.6   Reducing your risk of cardiovascular disease

44. _____ _____
45. _____ _____
46. _____ _____
47. _____
48. _____

51. _____
52. _____
53. _____
54. _____

49. _____
50. _____

55. _____
56. _____
57. _____

Figure 8.2

Fill-in-the-Blank

Referenced sections are indicated in parentheses.

1. Blood pressure is measured with a _____. (8.3)

2. In measuring blood pressure, the pressure at which blood reenters the collapsed artery is the _____ pressure. (8.3)

3. When diastolic pressure remains normal while systolic pressure is too high, this is called _____ _____ _____. (8.3)

4. _____ is a condition when blood pressure falls below normal. (8.3)

5. Cardiac output is determined by multiplying _____ _____ by _____ _____. (8.4)

6. _____ nerves from the _____ _____ of the brain can stimulate the heart to beat faster. (8.4)

7. Cardiovascular disorders are the number _____ killer in the United States. (8.5)

8. A(n) _____ is an X-ray of the blood vessels. (8.5)

9. _____ is a surgical procedure in which blood vessels are taken from one part of the body and grafted onto a blocked coronary artery. (8.5)

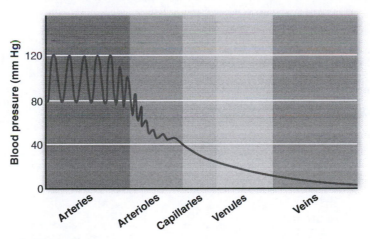

Figure 8.3

Short Answer

Answer the following questions about Figure 8.3.

10. a) What is represented by the highest peaks (120 mm Hg) in the arteries?

    b) What is happening in the heart during this time?

11. a) What is represented by the lowest peaks (80 mm Hg) in the arteries?

    b) What is happening in the heart during this time?

12. Why does blood pressure decrease in the arterioles and capillaries?

13. How does blood flow in the capillaries differ from blood flow in the arteries?

14. How do blood pressure differences in the arteries, capillaries, and veins keep blood flowing through the body?

True/False

Mark each statement as true or false. If a statement is marked false, replace the underlined term or phrase with the term or phrase needed to make the statement true.

_____ 15. Homeostatic regulation of the cardiovascular system depends on maintaining a constant <u>venous</u> blood pressure.

_____ 16. Baroreceptors are located in the <u>aorta and superior vena cava</u>.

_____ 17. Stretching of baroreceptors results in <u>increasing</u> cardiac output.

_____ 18. The heart beats more slowly when stimulated by <u>parasympathetic</u> nerves.

_____ 19. Increased metabolic activity in the tissues results in <u>dilation</u> of the precapillary sphincters.

Short Answer

For each cardiovascular system disorder listed below, indicate the cause and the symptoms.

20. Hypertension
    a. Cause:
    b. Symptoms:

21. Angina pectoris
    a. Cause:
    b. Symptoms:

22. Heart attack
    a. Cause:
    b. Symptoms:

23. Congestive heart failure
    a. Cause:
    b. Symptoms:

24. Embolism
    a. Cause:
    b. Symptoms:

25. Stroke
    a. Cause:
    b. Symptoms:

Looking Back/Looking Forward

1. Chapter 4 explained the components of negative feedback control systems. In the negative feedback system that regulates the cardiovascular system (described in section 8.4), identify the structure(s) serving as each of the following components:

   _____ a) controlled variable _____ c) control center

   _____ b) sensor _____ d) effector

2. One of the body's first defense mechanisms is an "inflammatory response" in which capillaries dilate, allowing immune cells to migrate into injured tissues. Why would dilation of capillaries increase their permeability?

Chapter Test

Word Choice

Circle the accurate term to complete each sentence.

1. Vasodilation of arterioles (increases/decreases) their diameter, (increasing/decreasing) blood flow to the capillaries.

2. The walls of veins are (thinner/thicker) than the walls of arteries.

3. The lumen of a vein is (larger/smaller) in diameter than the lumen of an artery.

4. Blood pressure in the veins is (less than/greater than) blood pressure in the arteries.

5. During diastole, pressure inside the heart chambers is (less than/greater than) blood pressure in the veins.

6. Baroreceptors stretch as blood pressure (increases/decreases).

7. Blood entering the right atrium is (oxygenated/deoxygenated).

Multiple Choice

8. The heart has _____ chambers and _____ valves.
   a. 2, 2
   b. 2, 4
   c. 4, 2
   d. 4, 4

9. In the functioning of the cardiovascular system, the heart provides
   _____, while the blood vessels provide
   _____.
   a. new red blood cells, oxygen
   b. baroreceptors to monitor blood pressure, smooth muscle capable of changing vessel diameter to respond to blood pressure
   c. power to move the blood, a network through which blood can flow
   d. hormones to regulate cardiac output, receptors to monitor cardiac output

10. Large- and medium-sized arteries are
   a. stiff, yet elastic.
   b. thin, yet strong.
   c. thick, yet diffusible.
   d. none of the above

11. Which layer of a large blood vessel's wall is primarily smooth muscle?
   a. inner layer
   b. middle layer
   c. outer layer
   d. epithelial layer

12. Which of the following accurately describes arterioles?
   a. They have the same outer layer of connective tissue as that found in arteries.
   b. They deliver blood directly to a venule.
   c. They help regulate the amount of blood that flows to each capillary.
   d. Blood pressure in the cardiovascular system is highest in the arterioles.

13. Capillary walls consist of
   a. a thin layer of smooth muscle surrounding the endothelium.
   b. smooth muscle enclosed by a thin layer of connective tissue.
   c. a single layer of endothelium composed of epithelial cells.
   d. three layers similar to larger arteries, but each layer is much thinner.

14. Excess extracellular fluid accumulates in the tissues because
   a. the capillary wall is too thick to allow the fluid to diffuse back into the vessels.
   b. the extracellular fluid contains too many proteins to allow return to the capillary.
   c. the pressure-induced filtration of water is greater than the diffusional reabsorption of water.
   d. the extracellular fluid provides storage for excess oxygen.

15. Homeostatic regulation of the cardiovascular system ensures that each tissue receives adequate blood flow. This process is most dependent on
   a. constant arterial blood pressure.
   b. constant venous blood pressure.
   c. ventricular systole.
   d. consistent heart rate.

16. Which of the following would interfere most with the ability of the veins to return blood to the heart?
    a. absence of valves in the veins
    b. low blood pressure in the veins
    c. high blood pressure in the arteries
    d. dilation of precapillary sphincters

17. The protective sac that encases the heart and anchors it to surrounding structures is the
    a. pericardium.
    b. endocardium.
    c. endothelium.
    d. myocardium.

18. Atrioventricular valves prevent blood from flowing backward into the
    a. atria.
    b. ventricles.
    c. pulmonary veins.
    d. aorta.

19. The movement of blood through the pulmonary circuit transports
    a. oxygenated blood from the lungs directly to the body.
    b. oxygenated blood from the lungs to the heart.
    c. deoxygenated blood from the lungs to the heart.
    d. deoxygenated blood from the left side of the heart to the right side of the heart.

20. The ventricles passively fill with blood during
    a. atrial systole.
    b. ventricular systole.
    c. atrial and ventricular diastole.
    d. semilunar diastole.

21. Which of the following represents the correct order in which electrical impulses will pass through the heart?
    a. AV node, AV bundle, SA node, Purkinje fibers
    b. SA node, Purkinje fibers, AV node, AV bundle
    c. AV bundle, AV node, SA node, Purkinje fibers
    d. SA node, AV node, AV bundle, Purkinje fibers

22. Which of the following would be occurring in a patient who lacked a P wave?
    a. no contraction of the atria
    b. no relaxation of the atria
    c. no contraction of the ventricles
    d. no relaxation of the ventricles

23. Shortness of breath, swollen ankles, and weight gain are symptoms of
    a. hypertension.
    b. hypotension.
    c. cerebral embolism.
    d. congestive heart failure.

24. An increase in cardiac output would result from all of the following *except*
    a. epinephrine.
    b. stimulation by sympathetic nerves.
    c. a decrease in stroke volume.
    d. an increase in heart rate.

25. If the cardiovascular center in the brain senses a drop in blood pressure, which of the following would occur?
    a. heart rate would be lowered
    b. cardiac output would be reduced
    c. vasoconstriction of the arterioles would occur
    d. parasympathetic nerves would carry signals to the heart

Self Test

Each of the Key Concepts listed at the beginning of this chapter has been rewritten as a question below. After successfully completing the study guide exercises and the Chapter Test, you should be able to answer each of these questions. Refer to the Key Concepts list to check your answers.

1. What provides the power to transport blood?

2. Which vessels transport blood away from the heart? Which vessels transport blood toward the heart?

3. What is the function of the capillaries?

4. What is important about the structure of blood vessel walls?

5. What three mechanisms help veins return blood to the heart?

6. How does the lymphatic system help to maintain blood volume?

7. How would you describe the four chambers of the heart?

8. What is the function of the heart valves?

9. What type of blood enters and then leaves each side of the heart?

10. What is the function of the pulmonary circuit?

11. What is the function of the systemic circuit?

12. How would you describe the periods of the cardiac cycle?

13. What is the role of the cardiac conduction system?

14. What is blood pressure?

15. Why is blood pressure important?

16. What produces systolic pressure? What produces diastolic pressure?

17. What three variables must be regulated to maintain a constant arterial blood pressure?

18. What factors contribute to regulation of the cardiovascular system?

19. What are the six disorders of the cardiovascular system discussed in Chapter 8?

20. How much effect can personal lifestyle choices have on cardiovascular health?

9

The Immune System and Mechanisms of Defense

Chapter Summary and Key Concepts

After reading and studying this chapter, you should know the following:

Sections 9.1, 9.2, 9.3, 9.4

1. The immune system is a group of cells, proteins, and structures of the lymphatic and circulatory systems that defend the body.

2. Pathogens are microorganisms that cause cell death and disease.

3. The danger of a particular pathogen is determined by its transmissibility, mode of transmission, and virulence.

4. The lymphatic system consists of a series of vessels, nodes, and organs that transport and cleanse the lymph.

5. The three lines of defense of the body are (1) keeping pathogens out, (2) nonspecific defense mechanisms, and (3) specific defense mechanisms.

6. The first line of defense is nonspecific, consisting of physical and chemical barriers, including skin, tears, saliva, earwax, digestive acids, mucus, vomiting, urination, defecation, and resident bacteria.

7. The second line of defense consists of nonspecific mechanisms that attack foreign and damaged cells, including phagocytosis by neutrophils and macrophages, destruction of large invaders by eosinophils, the inflammatory response, natural killer cells, the complement system, interferons, and fever.

Sections 9.5, 9.6, 9.7, 9.8

8. The third line of defense involves specific mechanisms that target specific pathogens for destruction, and depends on B cells and T cells.

9. The three important characteristics of the immune response are (1) it recognizes and targets specific pathogens, (2) it has memory, and (3) it protects the entire body.

10. An immune response is any response utilizing the components of the immune system. An antigen is any substance that can provoke an immune response.

11. Body cells are normally protected from the immune response by the presence of major histocompatibility complex (MHC) proteins.

12. Lymphocytes called B cells are responsible for antibody-mediated immunity. B cells produce antibodies capable of destroying a particular antigen.

13. Lymphocytes called T cells are responsible for cell-mediated immunity. T cells exist in several types and function to stimulate other cells and destroy infected body cells.

14. B cells recognize and bind to free antigens. T cells can only recognize and bind to an antigen that is presented by an antigen-presenting cell.

15. Both B cells and T cells produce memory cells that trigger the immune response faster when activated a second time. Memory cells are responsible for immunity.

16. Active immunization occurs when the body is exposed to a vaccine.

17. Passive immunization occurs when a person receives prepared antibodies for a specific pathogen.

18. Antibiotics are only effective against bacteria.

Sections 9.9, 9.10

19. Disorders of the immune system include allergies, lupus erythematosus, rheumatoid arthritis, and AIDS.

20. AIDS results from infection by the human immunodeficiency virus (HIV), and it causes a reduction in the number of helper T cells.

Exercises

Complete the exercises for each section after you have read and studied the section. If you cannot answer some questions, or if you answer them incorrectly, return to the chapter and review this information. You may find it helpful to work on only one section at a time. When you have completed all sections, take the Chapter Test as an indicator of your mastery of this topic.

9.1   Pathogens cause disease

9.2   The lymphatic system defends the body

True/False

Mark each statement as true or false. If you mark a statement as false, replace the underlined term or phrase with the term or phrase needed to make the statement true.

_____ 1. <u>All</u> bacteria are pathogenic.

_____ 2. <u>Viral</u> infections are treated with antibiotics.

_____   3. Bacteria are <u>eukaryotic</u> organisms.

_____   4. Bacteria are <u>larger</u> than viruses.

_____   5. Viruses are composed of a <u>single cell</u>.

_____   6. <u>Viruses</u> are responsible for causing Creutzfeldt-Jakob disease.

_____   7. <u>Prions</u> are a misfolded form of a normal brain protein.

_____   8. The most dangerous pathogens are <u>not easily transmitted</u> and have <u>low virulence</u>.

_____   9. The <u>red pulp</u> of the spleen contains macrophages.

_____   10. The thymus gland secretes <u>calcitonin and parathyroid hormone</u>.

Short Answer

11. List several organisms that may be pathogenic to humans.

12. Why is there disagreement among biologists concerning the characterization of viruses as living?

13. What factors determine the danger level of a particular pathogen?

14. List the three functions of the lymphatic system.

15. What are the components of the lymphatic system?

16. What are the two functions of the spleen and which part of the spleen is responsible for each function?

Labeling

Label the indicated structures in Figure 9.1. Indicate both (a) the structure and (b) the function.

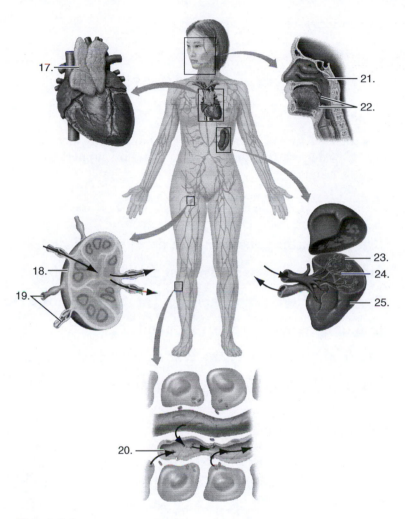

Figure 9.1

17. a. _____
    b. _____
18. a. _____
    b. _____
19. a. _____
    b. _____

20. a. _____
    b. _____
21. a. _____
    b. _____
22. a. _____
    b. _____

23. a. _____
    b. _____
24. a. _____
    b. _____
25. a. _____
    b. _____

Ordering

Beginning with a lymphatic capillary, indicate the order in which lymph will pass through the following structures as it is returned to the cardiovascular system.

lymphatic ducts   small lymphatic vessels   subclavian veins

larger lymphatic vessels   lymph nodes

26. Lymphatic capillary → a) _____ → b) _____ → c) _____
→ d) _____ → e) _____ → cardiovascular system

Critical Thinking

27. How is the structure of a lymphatic vessel different from the structure of an artery? How does the lymphatic vessel structure support the function of the lymphatic system?

9.3   **Keeping pathogens out: The first line of defense**

9.4   **Nonspecific defenses: The second line of defense**

9.5   **Specific defense mechanisms: The third line of defense**

9.6   **Immune memory creates immunity**

9.7   **Medical assistance in the war against pathogens**

9.8   **Tissue rejection: A medical challenge**

Crossword Puzzle

Across

1. T cells with CD8 receptors are activated to produce _____ T cells
3. Tissue _____ is minimized with the use of immunosuppresive drugs
7. Some activated B cells differentiate into _____ cells that produce antibodies
9. _____ antibodies are pure preparations specific for a single antigen
11. Self-markers on body cells are known as _____ histocompatability complex proteins
12. The region of an antibody where the amino acid sequence varies
13. A protein released by cytotoxic T cells that makes holes in the target cell membrane
14. Both B cells and cytotoxic T cells require stimulation by _____ T cells

Down

1. A class of signaling molecules secreted by helper T cells
2. Antigen-_____ cells display fragments of antigen on their surface
4. T cells are lymphocytes that mature in this gland
5. A first exposure to a particular antigen generates a _____ immune response
6. Antibiotics are effective against these microorganisms
8. B cells are responsible for _____-mediated immunity
10. Active immunization occurs when this is administered
11. The presence of these cells is the basis of immunity

Fill-in-the-Blank

Referenced sections are in parentheses.

15. Tears and saliva contain _____ an enzyme that kills bacteria. (9.3)

16. Some resident bacteria are _____ because they compete with pathogenic bacteria for resources. (9.3)

17. Mast cells secrete _____ to promote vasodilation of small blood vessels. (9.4)

18. _____ is a group of proteins secreted by virus-infected cells. (9.4)

19. When attacking bacteria, macrophages can release _____ into the bloodstream to reset the body's thermostat and cause a fever. (9.4)

20. A(n) _____ is any substance that provokes an immune response. (9.5)

Short Answer

21. For each cell listed below, write which line of defense (first, second, or third) includes the cell, and the function of each cell.
    a. neutrophils

    b. macrophages

    c. eosinophils

    d. basophils

    e. natural killer cells

    f. B cells

    g. cytotoxic T cells

22. Your friend has stepped on a nail, and is complaining about the pain at the injury site, which is now reddened, swollen, and warm to the touch. Explain to your friend what is happening and why these events and symptoms are actually positive developments.

Completion

Write the name of each type of antibody described in the table below. Indicate what percentage of immunoglobulins each antibody type comprises.

Type of Antibody	% of immunoglobulins	Antibody Function
23. IgA	a.	b.
24. IgE	a.	b.
25. IgG	a.	b.
26. IgD	a.	b.
27. IgM	a.	b.

9.9   Inappropriate immune system activity causes problems

9.10   Immune deficiency: The special case of AIDS

Fill-in-the-Blank

Referenced sections are in parentheses.

1. _____ is an inappropriate response of the immune system to a nondangerous pathogen. (9.9)

2. Anaphylactic shock is a severe _____ allergic reaction. (9.9)

3. _____ disorders result when the immune system targets "self" cells. (9.9)

4. _____ is an autoimmune disorder in which the body attacks its own connective tissue. (9.9)

Completion

Complete each column of the table.

Phase	Developmental Timeline	Symptoms	T-cell Level	Disease Status
5. Phase I	a)	b)	c)	d)
6. Phase II	a)	b)	c)	d)
7. Phase III	a)	b)	c)	d)

Looking Back/Looking Forward

1. Antibiotics interfere with bacterial reproduction in several different ways. Would it be effective to market an antibiotic that stopped bacterial reproduction by interfering with the production of the plasma membrane? Explain your answer.

2. The digestive tract, discussed in Chapter 14, includes a series of organs and passageways from the mouth to the anus. Describe how the body is protected at various points along the digestive tract.

Chapter Test

Multiple Choice

1. Microorganisms that cause cell death and disease are called
   a. antigens.
   b. antibodies.
   c. macrophages.
   d. pathogens.

2. Characteristics of bacteria include all of the following *except*
   a. presence of membrane-bound organelles.
   b. a single circular chromosome.
   c. absence of a nuclear membrane.
   d. presence of a cell wall.

3. Viruses
   a. are independent, single-celled organisms.
   b. must use host cell machinery to reproduce.
   c. are responsible for vCJD.
   d. produce prions.

4. Which of the following is not a characteristic shared by both lymphatic vessels and veins?
   a. walls consisting of three layers
   b. use skeletal muscle contractions to keep fluids moving.
   c. one-way valves
   d. fluid flows through lymph nodes

5. The correct path of lymph fluid collected from the tissues and transported to the heart is most accurately represented as
   a. lymphatic duct → lymph node → lymphatic capillary → lymphatic vessel.
   b. lymphatic vessel → lymphatic capillary → lymphatic duct → lymph node.

    c. lymph node → lymphatic capillary → lymphatic vessel → lymphatic duct.

    d. lymphatic capillary → lymphatic vessel → lymph node → lymphatic duct.

6. Lymph nodes are clustered in all of the following areas of the body *except* the
   a. neck.
   b. groin.
   c. ankles.
   d. digestive tract.

7. The spleen
   a. is the largest lymphatic organ.
   b. includes white pulp that contains macrophages.
   c. cleanses the lymph.
   d. is required to maintain life and health.

8. Specific defense mechanisms differ from nonspecific defense mechanisms because only specific mechanisms
   a. utilize macrophages.
   b. target particular antigens.
   c. protect a localized region.
   d. involve complement proteins.

9. What do lysozyme, dermicidin, and mucus have in common?
   a. They are all produced by lymphocytes.
   b. They all circulate in lymphatic fluid.
   c. They are all part of the first line of defense.
   d. They are all associated with the action of macrophages.

10. The first white blood cells to respond to an infection are
    a. eosinophils.
    b. macrophages.
    c. neutrophils.
    d. helper B cells.

11. A group of at least 20 proteins that circulate in the blood and assist defense mechanisms is called
    a. natural killer cells.
    b. interferons.
    c. complement.
    d. cytokines.

12. Fever is caused when _____ release _____.
    a. macrophages, pyrogens
    b. infected cells, interferon
    c. B cells, cytokines
    d. eosinophils, histamines

13. A substance capable of provoking an immune response is a(n)
    a. antibody.
    b. antigen.

    c.   lymphokine.

    d.   complement.

14. Major histocompatibility complex proteins
    a. are found in the cytoplasm.
    b. attract pathogens.
    c. identify a cell as "self."
    d. are produced by B cells.

15. B cells mature in the _____, and T cells mature in the
    _____.
    a. bone marrow, bone marrow
    b. bone marrow, thymus gland
    c. thymus gland, bone marrow
    d. thymus gland, thymus gland

16. Specific immunity is dependent on the body's ability to
    a. distinguish its own cells from those of foreign invaders.
    b. manufacture adequate numbers of red blood cells.
    c. maintain a first and second line of defense.
    d. circulate and replenish lymphatic fluid.

17. B cells are responsible for _____ immunity, and T cells are
    responsible for _____ immunity.
    a. cell-mediated, antibody-mediated
    b. antibody-mediated, cell-mediated
    c. specific, nonspecific
    d. active, passive

18. When a T cell with CD8 receptors contacts an APC displaying an antigen-MHC complex,
    a. the APC divides to form helper T cells.
    b. the T cell produces a clone of cytotoxic T cells.
    c. the antigen-MHC complex is destroyed.
    d. the CD8 receptors are converted to CD4 receptors.

19. A specific antibody will bind to
    a. a general class of antigens.
    b. a specific antigen.
    c. any antigen, because the heavy chains can accommodate any shape.
    d. any antigen, because the binding sites are variable.

20. Which of the following can function as antigen-presenting cells?
    a. natural killer cells
    b. macrophages
    c. mast cells
    d. antibodies

21. A secondary immune response could not occur without
    a. memory cells.
    b. complement.
    c. neutrophils.
    d. natural killer cells.

22. Administration of prepared antibodies produces
    a. active immunity.
    b. passive immunity.
    c. delayed immunity.
    d. permanent immunity.

23. The actions and effectiveness of antibiotics depends on differences between bacterial and human cells, including the following:
    a. Bacteria have a thick cell wall, human cells do not.
    b. The membrane-bound organelles of bacteria contain different proteins than those of human cells.
    c. Bacteria do not have ribosomes.
    d. Human protein synthesis occurs faster than bacterial protein synthesis.

24. HIV interferes with specific immunity because it infects and eventually destroys
    a. B cells.
    b. cytotoxic T cells.
    c. helper T cells.
    d. MHC cells.

25. Autoimmune disorders occur when
    a. the immune system loses a large number of T cells.
    b. the ability of B cells to produce antibodies is reduced.
    c. the immune system fails to recognize "self" cells.
    d. macrophages function only in nonspecific immunity.

Self Test

Each of the Key Concepts listed at the beginning of this chapter has been rewritten as a question below. After successfully completing the study guide exercises and the Chapter Test, you should be able to answer each of these questions. Refer to the Key Concepts list at the beginning of this chapter to check your answers.

1. The immune system comprises which molecules, structures, and systems?

2. What are pathogens?

3. What determines the danger of a particular pathogen?

4. The lymphatic system comprises which structures and organs?

5. What are the three lines of defense of the body?

6. Which structures and processes make up the first line of defense?

7. How would you describe the structures and processes that make up the second line of defense?

8. Is the third line of defense specific or nonspecific? What cells are responsible for the third line of defense?

9. What are the three important characteristics of the immune response?

10. What is an immune response? What is an antigen?

11. How are body cells normally protected from the immune system?

12. What type of immunity involves B cells? What molecules do B cells produce?

13. What type of immunity involves T cells? What do T cells do?

14. What type of antigen is recognized by B cells? What type of antigen is recognized by T cells?

15. What process are memory cells responsible for? What cells produce memory cells?

16. What results in active immunity?

17. What results in passive immunity?

18. What pathogens are antibiotics effective against?

19. What are four disorders of the immune system discussed in the chapter?

20. What causes AIDS? What effect does AIDS have on T cells?

10

The Respiratory System: Exchange of Gases

Chapter Summary and Key Concepts

After reading and studying this chapter, you should know the following:

Sections 10.1, 10.2

1. The term respiration includes four processes: ventilation, external respiration, internal respiration, and cellular respiration.

2. Ventilation is the process of moving air into and out of the lungs, external respiration is the exchange of gases between inhaled air and the blood, internal respiration is the exchange of gases between the blood and tissue fluids, and cellular respiration is the process of using oxygen to produce ATP within cells.

3. The upper respiratory tract consists of the nose and pharynx, and it functions to filter, warm, and humidify air.

4. The lower respiratory tract consists of the larynx, trachea, bronchi, and lungs, and it functions to move air and exchange gases.

5. The pharynx is a common passageway for food, liquid, and air. Food and liquid are prevented from moving into the air passageways by the epiglottis.

6. The larynx also functions as the voice box. Vibration of the vocal cords produces sound.

7. The trachea is reinforced with incomplete cartilage rings; it branches into the bronchi, which are reinforced by cartilage tissue; the bronchi further branch into the bronchioles, which have no cartilage.

8. Bronchioles terminate in tiny air-filled sacs called alveoli that function as the site of gas exchange.

9. The lungs are in the thoracic cavity and are enclosed by the pleural membranes.

Sections 10.3, 10.4, 10.5, 10.6

10. Contraction of the intercostal muscles and the diaphragm enlarges the thoracic cavity, resulting in enlargement of the lungs.

11. Air pressure within the lungs decreases when lung volume increases.

12. Air moves into the lungs down a pressure gradient.

13. Inspiration is the movement of air into the lungs. It is an active process because it requires muscular effort.

14. Passive relaxation of the intercostals and the diaphragm reduces lung volume, increasing air pressure within the lungs. As the air pressure within the lungs becomes greater than the atmospheric pressure, air rushes out of the lungs.

15. Expiration is the movement of air out of the lungs. It is usually a passive process.

16. Tidal volume, vital capacity, inspiratory reserve volume, expiratory reserve volume, and residual volume are measurements of lung capacity and can reflect lung function.

17. External respiration and internal respiration occur when gases move passively along a pressure gradient.

18. Most oxygen is transported through the blood bound to hemoglobin. Carbon dioxide is transported through the blood bound to hemoglobin or in the form of a bicarbonate ion.

19. Breathing is regulated by the nervous system.

20. Disorders of the respiratory system include asthma, emphysema, bronchitis, pneumonia, tuberculosis, botulism, cancer, and congestive heart failure.

Exercises

Complete the exercises for each section after you have read and studied the section. If you cannot answer some questions, or if you answer them incorrectly, return to the chapter and review this information. You may find it helpful to work on only one section at a time. When you have completed all sections, take the Chapter Test as an indicator of your mastery of this topic.

10.1 Respiration takes place throughout the body

10.2 The respiratory system consists of upper and lower respiratory tracts

Define

Write the term(s) for each definition.

_____ 1. the organ system responsible for the exchange of oxygen and carbon dioxide between the body and the air

_____ 2. another term for breathing

_____ 3. the exchange of gases between inhaled air and the blood

_____ 4. the exchange of gases between the blood and the tissue fluids

_____ 5. the process of using oxygen to produce ATP within cells

_____ 6. the visible portion of the nose

_____   7. the internal portion of the nose

_____   8. epithelial tissue lining the nasal cavity is covered with these small hair-like projections

_____   9. air spaces inside the skull

_____   10. the voice box

_____   11. flap of cartilage at the opening to the voice box

_____   12. the opening to the airway, surrounded by the vocal cords

_____   13. the section of the respiratory tract that consists of C-shaped rings of cartilage held together by connective tissue and muscle

Fill-in-the-Blank

Referenced sections are in parentheses.

14. The four processes included in respiration are _____,

_____, _____, and _____. (10.1)

15. The exchange of gases between the blood and tissue fluids is _____ respiration. (10.1)

16. The process of using oxygen to produce ATP within cells is _____ respiration. (10.1)

17. In addition to gas exchange, the respiratory system is also responsible for the production of _____. (10.1)

18. The upper respiratory tract consists of the _____ and the _____. (10.2)

19. The lower respiratory tract consists of the _____, _____, _____, and _____. (10.2)

20. The two chambers of the nose are separated by the _____ _____. (10.2)

21. Gas exchange between the air in the alveoli and the blood in the pulmonary capillaries occurs by _____. (10.2)

22. Both food and air pass through the _____. (10.2)

23. The pharynx connects the mouth and nasal cavity to the _____. (10.2)

24. _____ are air spaces in the skull that secrete mucus and drain into the nasal cavity. (10.2)

25. The larynx contains the _____ and the _____. (10.2)

26. The _____ are marked by the absence of cartilage. (10.2)

27. Friction in the pleural membrane during lung movement is reduced by the presence of

_____ in the _____ _____. (10.2)

28. The right lung has _____ lobes and the left lung has

_____ lobes. (10.3)

29. The wall of a single alveolus is _____ layer(s) of _____

_____ cells. (10.2)

30. Surface tension within the alveoli is reduced by _____. (10.2)

31. Gas exchange in the lungs occurs between the _____ and the

_____ _____. (10.2)

Labeling/Coloring

Select different colors for each structure listed below. Color the coding circles and the corresponding structures in Figure 10.1.

- ◯ nose
- ◯ mouth
- ◯ epiglottis
- ◯ pleural membranes
- ◯ lung
- ◯ intercostal muscle
- ◯ rib
- ◯ alveoli
- ◯ diaphram
- ◯ nasal cavity
- ◯ pharynx
- ◯ trachea
- ◯ bronchi
- ◯ larynx

Figure 10.1

10.3 The process of breathing involves a pressure gradient

Fill-in-the-Blank

1. Gas pressure is caused when gas molecules _____.

2. When the volume of a closed space increases, the gas pressure

   _____.

3. Gases flow from an area of _____ pressure to an area of

   _____ pressure.

4. Inspiration and expiration occur when air moves down its

   _____ _____.

5. The intercostal muscles and the _____ are involved in

   moving air into and out of the lungs.

Word Choice

The following statements refer to the cycle of inspiration and expiration. Circle the accurate term or phrase to complete each sentence.

6. At rest, the intercostal muscles are relaxed and the diaphragm is (relaxed/contracted).

7. During inspiration
   a. the diaphragm (relaxes/contracts).
   b. the volume of the pleural cavity (increases/decreases).
   c. the volume of the lungs (increases/decreases).
   d. the pressure within the pleural space (increases/decreases).

8. When the air pressure in the lungs (increases/decreases), air will rush in.

9. During expiration
   a. the diaphragm becomes (flattened/dome shaped).
   b. the ribs move (upward/downward).
   c. the lungs become (larger/smaller).
   d. the air pressure in the lungs (increases/decreases).

10. During quiet breathing
    a. inspiration is (active/passive/may be passive or active).
    b. expiration is (active/passive/may be passive or active).

11. During exercise
    a. inspiration is (active/passive/may be passive or active).
    b. expiration is (active/passive/may be passive or active).

Labeling

Write the definition of each term listed below, and then label Figure 10.2 with the correct term.

12. Expiratory Reserve:

13. Vital Capacity:

14. Tidal Volume:

15. Residual Volume:

16. Inspiratory Reserve:

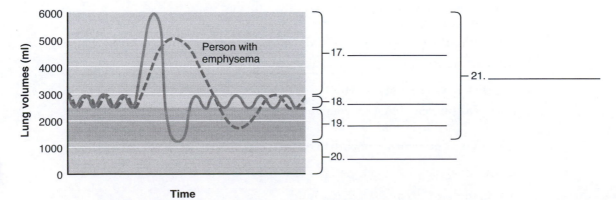

Figure 10.2

10.4   Gas exchange and transport occurs passively

Fill-in-the-Blank

Refer to Figure 10.11 in your text as you answer questions 1 and 2. For questions 3, 4, and 5, formulas are included that represent a process involved in the transportation of O_2 or CO_2 in the blood; refer to these formulas as you answer the questions.

1. Gas exchange between the alveoli and the pulmonary capillaries:

    a. What is the PO_2 in the alveoli? _____

    b. What is the PO_2 in the pulmonary capillaries? _____

    c. What direction will O_2 move? _____

    d. What is the PCO_2 in the alveoli? _____

    e. What is the PCO_2 in the pulmonary capillaries? _____

    f. What direction will CO_2 move? _____

    g. What type of respiration is this? _____

2. Gas exchange between the capillary blood and the interstitial fluid:

    a. What is the PO_2 in the capillary blood? _____

    b. What is the PO_2 in a cell? _____

    c. What direction will O_2 move? _____

    d. What is the PCO_2 in the capillary blood? _____

    e. What is the PCO_2 in a cell? _____

    f. What direction will CO_2 move? _____

    g. What type of respiration is this? _____

3. $Hb + O_2 \rightarrow HbO_2$

    a. HbO_2 is a molecule of _____ bound to _____ and is called _____.

    b. This reaction is _____.

    c. List three conditions that promote the binding of O_2 to hemoglobin: _____, _____, and _____.

4. $Hb + CO_2 \rightarrow HbCO_2$

    a. $HbCO_2$ is a molecule of _____ bound to _____ _____ and is called _____.

    b. What percentage of carbon dioxide is transported in the blood this way? _____

5. $CO_2 + H_2O \rightarrow H_2CO_3 \rightarrow HCO_3 + H^+$

    a. In this reaction, CO_2 combines with water in the blood to produce a _____ ion.

    b. What percentage of carbon dioxide is transported in the blood this way? _____

10.5 The nervous system regulates breathing

10.6 Disorders of the respiratory system

Paragraph Completion

Complete the paragraphs below by writing in the missing term(s).

Respiration is controlled by the 1) _____ system. Electrical impulses that control breathing are generated automatically every 2) _____ seconds in the 3) _____ center of the brain, located in the 4) _____. These impulses travel to the diaphragm and 5) _____ muscles. As these muscles contract, the rib cage 6) _____ as the diaphragm is pulled 7) _____, causing us to 8) _____. 9) _____ receptors in the lungs monitor the degree of inflation and send this information to the 10) _____ center, limiting inhalation and eventually triggering 11) _____. When impulses from the respiratory center end, the muscles 12) _____, the rib cage returns to its original size, the diaphragm moves 13) _____, and we 14) _____.

The regulation of breathing centers on maintaining homeostasis of 15) _____, 16) _____, and 17) _____. Levels of 18) _____ are most critical to maintaining homeostasis. When CO_2 levels rise as a normal effect of internal respiration, this results in a rise in 19) _____ concentration as CO_2 combines with H_2O in the blood. Changes in CO_2 levels are detected by monitoring the level of 20) _____ in the 21) _____ fluid. When these H^+ levels rise, the respiratory center stimulates 22) _____ breathing, causing us to exhale more 23) _____. A change in blood CO_2 levels of 24) ____ % is sufficient to stimulate a response.

Some receptors monitor 25) _____ levels rather than CO_2 levels. These receptors are the 26) _____ bodies located in the 27) _____. If O_2 levels in the blood fall, these receptors signal the 28) _____ center to 29) _____ the rate and depth of breathing. The carotid and aortic bodies are only stimulated if blood O_2 levels fall by at least 30) _____ %.

Critical Thinking

31. A fall or injury involving a blow to the base of the skull, the area where the medulla oblongata is located, can result in death. Explain why such an injury might be fatal.

32. Regulation of breathing depends largely on CO_2 levels, which are monitored indirectly by cells that detect changes in H^+ concentrations in the cerebrospinal fluid. Why would it be ineffective to have H^+ levels monitored by cells in the carotid and aortic bodies?

Short Answer

For each respiratory system disorder listed below, indicate the cause and the symptoms.

33. Asthma

    a. Cause:

    b. Symptoms:

34. Emphysema

    a. Cause:

    b. Symptoms:

35. Bronchitis

    a. Cause:

    b. Symptoms:

36. Pneumonia

    a. Cause:

    b. Symptoms:

37. Tuberculosis

    a. Cause:

    b. Symptoms:

38. Botulism

    a. Cause:

    b. Symptoms:

39. Lung cancer

    a. Cause:

    b. Symptoms:

40. Cystic fibrosis

    a. Cause:

    b. Symptoms:

41. Congestive heart failure

    a. Cause:

    b. Symptoms:

42. Pneumothorax

    a. Cause:

    b. Symptoms:

Looking Back/Looking Forward

1. The alveoli contain macrophages, whose function was presented in Chapter 9. Why is this an important location for macrophages?

2. Chapter 21 discusses aging. As aging progresses, the alveoli enlarge and atrophy, the ribs become less flexible, and the chest wall loses some ability to expand. What effect will these changes have on vital capacity and residual volume in the elderly?

Chapter Test

Multiple Choice

1. Which of the following terms is matched with the incorrect definition?
   a. ventilation—the movement of air into and out of the lungs
   b. internal respiration—the exchange of gases between the blood and the tissue fluids
   c. external respiration—the exchange of gases between inhaled air and the blood
   d. cellular respiration—the exchange of gases between the tissue fluids and the cells

2. All of the following are part of the lower respiratory tract *except* the
   a. larynx.
   b. pharynx.
   c. trachea.
   d. bronchi.

3. Cilia line the
   a. nasal cavity.
   b. pharynx.
   c. trachea.
   d. bronchioles.
   e. a and c
   f. all of the above

4. Which of the following is *not* a function of the nose?
   a. receptors provide for the sense of smell
   b. gas exchange
   c. filtering inhaled air
   d. moistening air

5. When air flows through the larynx,
   a. both the epiglottis and the glottis are open.
   b. the epiglottis is closed and the glottis is open.
   c. the epiglottis is open and the glottis closed.
   d. both the epiglottis and the glottis are closed.

6. Which of the following structures that influence vocalization are incorrectly matched with their function?
   a. tongue and soft palate—create recognizable sounds
   b. pharynx and nose—act as resonating chambers to amplify and enhance vocal tone
   c. vocal cords—vibrate to produce sound
   d. nasal sinuses—push air through the vocal cords for sound production

7. The incomplete cartilage rings of the trachea allow the trachea to
   a. maintain a constant diameter at all times.
   b. push air downward with muscle movements.
   c. close when needed to prevent choking.
   d. change diameter to accommodate coughing or heavy breathing.

8. The alveoli
   a. consist of a single layer of squamous epithelial cells.
   b. are reinforced by cartilage tissue.
   c. do not participate in gas exchange.
   d. fuse to form the bronchioles.

9. The lungs are enclosed by a double pleural membrane. These membranes are attached to the _____ and the _____.
   a. thoracic cavity, pleural cavity
   b. pleural cavity, lung surface
   c. thoracic cavity, lung surface
   d. thoracic cavity, abdominal cavity

10. Gas exchange between inhaled air and the blood is accomplished by the _____ and the _____.
    a. alveoli, pulmonary veins
    b. bronchioles, pulmonary capillaries
    c. alveoli, pulmonary capillaries
    d. alveoli, bronchioles

11. Gases flow passively from an area of _____ pressure to an area of _____ pressure.
    a. high, low
    b. low, high
    c. high, high
    d. low, low

12. Air moves into the lungs when the air pressure in the lungs is
    _____ the air pressure in the atmosphere.
    a. lower than
    b. higher than
    c. the same as
    d. Differences in air pressure are not responsible for airflow.

13. Expiration is
    a. always a passive process.
    b. always an active process.
    c. usually passive, but it can be active.
    d. usually active, but it can be passive.

14. The air that always remains in the lungs is referred to as the
    a. tidal volume.
    b. vital capacity.
    c. inspiratory reserve volume.
    d. expiratory reserve volume.
    e. residual volume.

15. Measurements of lung capacity are performed with a device called a(n)
    a. volumetric analyzer.
    b. spirometer.
    c. inhalizer.
    d. airflow compression chamber.

16. Under which of the following conditions would oxygen flow from the alveoli
    into the blood?
    a. PO_2 in the alveoli is 40, PO_2 in the blood is 100
    b. PO_2 in the alveoli is 104, PO_2 in the blood is 104
    c. PCO_2 in the alveoli is 40, PO_2 in the alveoli is 46
    d. PO_2 in the alveoli is 104, PO_2 in the blood is 40

17. The earth's atmosphere consists primarily of three gases, which are
    a. hydrogen, oxygen, and carbon.
    b. nitrogen, oxygen, and carbon dioxide.
    c. nitrogen, oxygen, and water.
    d. oxygen, hydrogen, and carbon dioxide.

18. Venous blood is defined as blood that is
    a. found in the veins.
    b. low in carbon dioxide.
    c. traveling toward the heart.
    d. deoxygenated.

19. Hemoglobin in the blood transports
    a. oxygen only.
    b. carbon dioxide only.
    c. oxygen and carbon dioxide in equal amounts.
    d. more oxygen than carbon dioxide.
    e. more carbon dioxide than oxygen.

20. Most of the carbon dioxide transported in the blood to the lungs will be transported
    a. by hemoglobin.
    b. by red blood cells.
    c. as bicarbonate ions in the red blood cells.
    d. as bicarbonate ions in the plasma.

21. A low pH reduces the binding affinity of hemoglobin for oxygen, increasing the release of oxygen and its delivery to the tissues. What produces a condition of low pH in the red blood cells?
    a. conversion of CO_2 to bicarbonate ions
    b. conversion of O_2 to oxyhemoglobin
    c. conversion of CO_2 to carbaminohemoglobin
    d. conversion of bicarbonate ions to CO_2

22. The respiratory center
    a. monitors the level of oxygen in the blood.
    b. contains stretch receptors that limit lung expansion through negative feedback.
    c. controls the rate of breathing.
    d. is the site of gas exchange between the air and the blood.

23. The concentration of H^+ in the cerebrospinal fluid rises when
    a. the PO_2 of the arterial blood rises.
    b. the PO_2 of the arterial blood falls.
    c. the PCO_2 of the arterial blood rises.
    d. the PCO_2 of the arterial blood falls.

24. The carotid and aortic bodies are arterial receptors that respond to
    a. a decrease in the PO_2 of arterial blood.
    b. an increase in the PO_2 of arterial blood.
    c. a decrease in the PO_2 of venous blood.
    d. an increase in the PO_2 of venous blood.

25. Which of the following disorders of the respiratory system reduces the surface area available for gas exchange in the lungs?
    a. asthma
    b. bronchitis
    c. emphysema
    d. pneumonia

Self Test

Each of the Key Concepts listed at the beginning of this chapter has been rewritten as a question below. After successfully completing the study guide exercises and the Chapter Test, you should be able to answer each of these questions. Refer to the Key Concepts list at the beginning of this chapter to check your answers.

1. What are the four processes encompassed by the term respiration?

2. Describe what occurs in each of the four processes of respiration.

3. What are the components and function of the upper respiratory tract?

4. What are the components and function of the lower respiratory tract?

5. What acts as a common passageway for food, liquid, and air? What prevents food and liquid from moving into the airways?

6. Where and how is sound produced?

7. How would you compare the amount of cartilage present in the trachea, bronchi, and bronchioles?

8. What structures are located at the end of the bronchioles? What process occurs here?

9. What body cavity contains the lungs? What membrane encloses the lungs?

10. What are two skeletal muscles that contribute to lung function?

11. What happens to the air pressure in the lungs as the lung volume increases?

12. How would you compare the air pressure in the atmosphere with air pressure in the lungs when air flows into the lungs?

13. What is inspiration? Why is it considered an active process?

14. What muscle activity reduces lung volume? Compare the air pressure in the atmosphere with the air pressure in the lungs when air flows out of the lungs.

15. What is expiration? Is it active or passive?

16. What are five measurements of lung capacity?

17. Do gases move with or against a pressure gradient during external and internal respiration?

18. How is most oxygen transported in the blood? How is carbon dioxide transported in the blood?

19. What body system regulates breathing?

20. What are the eight disorders of the respiratory system discussed in the chapter?

11

The Nervous System: Integration and Control

Chapter Summary and Key Concepts

After reading and studying this chapter, you should know the following:

Sections 11.1, 11.2, 11.3, 11.4, 11.5

1. The three characteristics of the nervous system are (1) it requires information, (2) it integrates information, and (3) it can respond to information very quickly.

2. The two principal parts of the nervous system are the central nervous system (CNS), consisting of the brain and spinal cord, and the peripheral nervous system (PNS), consisting of all nervous tissue not in the CNS.

3. Neurons are nervous system cells specialized to receive and transmit information.

4. The three classifications of neurons are (1) sensory neurons that carry information to the CNS, (2) motor neurons that carry information away from the CNS, and (3) interneurons that communicate between nervous system components.

5. The parts of a neuron are the cell body, dendrites, axon hillock, axon, myelin sheath, nodes of Ranvier, axon terminals, and axon bulbs.

6. The slight difference between the voltage inside and outside a cell is the resting membrane potential.

7. A graded potential occurs when chemically sensitive channels allow ion movements that depolarize or hyperpolarize the membrane.

8. An action potential is a reversal in the voltage difference across a membrane and is triggered when the sum of all graded potentials reaches a threshold level.

9. The three events of an action potential are depolarization, repolarization, and reestablishment of the resting potential.

10. During an action potential, depolarization occurs when sodium moves into the cell, repolarization occurs when potassium moves out of the cell, and reestablishment of the resting membrane potential is accomplished by the sodium-potassium pump.

11. Neurotransmitters are chemicals that carry information across a synapse from one neuron to a target cell. Neurotransmitters that are excitatory depolarize the postsynaptic cell, and neurotransmitters that are inhibitory hyperpolarize the postsynaptic cell.

12. Postsynaptic cells integrate and process information.

Sections 11.6, 11.7, 11.8, 11.9, 11.10, 11.11, 11.12, 11.13

13. A nerve contains many axons bundled together. Cranial nerves attach to the brain, and spinal nerves attach to the spinal cord.

14. The PNS is divided into the sensory nervous system that carries information to the CNS, and the motor nervous system that carries information away from the CNS.

15. The motor nervous system branches into the somatic division that controls skeletal muscle, and the autonomic division that controls smooth muscle, cardiac muscle, and glands. The autonomic division branches further into the sympathetic and parasympathetic nervous systems.

16. The three regions of the brain are the hindbrain, which includes the medulla oblongata, pons, and cerebellum; the forebrain, which includes the thalamus, hypothalamus, limbic system, and cerebrum; and the midbrain.

17. The cerebrum is responsible for higher functions and is divided into the occipital, temporal, parietal, and frontal lobes.

18. The reticular activating system is responsible for levels of sleep and wakefulness. The limbic system in the forebrain is the site of emotion and basic behaviors.

19. The limbic system manages short-term memories for a few hours. Long-term memories occur in the cerebrum and may last for years.

20. Psychoactive drugs affect higher brain functions and can be addictive. Disorders of the nervous system include trauma, infections, abnormal neural activity, and tumors.

Exercises

Complete the exercises for each section after you have read and studied the section. If you cannot answer some questions, or if you answer them incorrectly, return to the chapter and review this information. You may find it helpful to work on only one section at a time. When you have completed all sections, take the Chapter Test as an indicator of your mastery of this topic.

11.1   The nervous system has two principal parts

11.2   Neurons are the communication cells of the nervous system

11.3   Neurons initiate action potentials

11.4   Neuroglial cells support and protect neurons

11.5   Information is transferred from a neuron to its target

Define

Write the term(s) for each definition.

_____ 1. the organ system that receives, integrates, and responds to information

_____ 2. nervous system cells that transmit information in the form of electrical impulses

_____ 3. the main body of a neuron

_____ 4. extensions of a neuron that receive information

_____ 5. a long, slender extension of a neuron that transmits information by conducting an electrical impulse

_____ 6. area of a motor neuron where an electrical impulse is generated

_____ 7. a change in membrane potential that moves the voltage closer to zero

_____ 8. a change in membrane potential that makes the voltage more negative

_____ 9. transient local changes in the resting membrane potential

_____ 10. when many graded potentials produce a bigger change in the membrane potential than one impulse alone

_____ 11. nervous system cells that provide support to neurons

_____ 12. cells of the peripheral nervous system that produce a myelin sheath

_____ 13. the "leap-frog" conduction pattern that occurs along myelinated neurons

_____ 14. cells of the central nervous system that produce a myelin sheath

_____ 15. a specialized junction between two cells where information is transmitted by a neurotransmitter

_____ 16. small gap that separates a presynaptic and postsynaptic membrane

Short Answer

17. What are the two branches of the nervous system, and what is comprised in each?

18. What are the two functional branches of the PNS and what is the function of each?

19. What are the two functional branches of the motor division of the PNS, and what is the function of each?

20. What are the two subdivisions of the autonomic branch, and how do they work with each other?

21. List the three types of neurons and describe the function of each.

22. How does the sodium-potassium pump contribute to the resting membrane potential?

23. What are the three functions of the myelin sheath of PNS neurons?

Diagram

Draw a motor neuron in the space provided. Include the following components and label them in your diagram.

axon	**cell body**	**axon terminal**	**axon bulb**
dendrite	**Schwann cell**	**node of Ranvier**	**myelin sheath**

Labeling

Use the following terms to label each indicated area of Figure 11.1. Then, write a description of each term on the lines provided.

Hyperpolarizing graded potential:

_____

_____

Threshold:

_____

_____

Resting membrane potential:

_____

_____

Summation:

_____

_____

Depolarizing graded potential:

_____

_____

Action potential:

_____

_____

Figure 11.1

Word Choice

Circle the term or phrase that correctly completes each sentence.

30. Graded potentials (are/are not) self-propagating.

31. The opening of sodium channels on the axon allows Na^+ to move (into/out of) the cell.

32. An action potential (is/is not) self-propagating.

33. The speed of an action potential increases as the diameter of the axon (increases/decreases).

34. When an action potential arrives at the axon bulb, calcium channels (open/close) and Ca^{++} moves (into/out of) the bulb.

35. The effect of a neurotransmitter is relatively (short-lived/long-lived).

36. Muscle cells (do/do not) process and integrate information, because they receive input from (one/multiple) neuron(s).

37. Excitatory neurotransmitters (depolarize/repolarize) the (presynaptic/postsynaptic) cell.

38. At a synapse, neurotransmitters will be released by the (presynaptic/postsynaptic) membrane.

39. When molecules of neurotransmitter bind receptors on a postsynaptic membrane (chemically sensitive/voltage sensitive) channels will open.

40. Integration of information occurs in the (presynaptic/postsynaptic) cell.

Ordering

Place the following events associated with an action potential in the correct order of occurrence, beginning with the graded potential reaching threshold.

a) graded potential reaches threshold

b) K^+ diffuses out of the axon

c) axon is slightly hyperpolarized

d) voltage-sensitive Na^+ channels open on the axon

e) repolarization of the axon occurs

f) resting potential is reestablished

g) Na^+ diffuses into the cytoplasm of the axon

h) K^+ channels close

i) depolarization of the axon occurs

j) Na^+ channels close and K+ channels open

41. (a) → ＿＿ → ＿＿ → ＿＿ → ＿＿ → ＿＿ → ＿＿ → ＿＿ → ＿＿ → ＿＿

Critical Thinking

42. If the sodium-potassium pump failed to transport Na^+ out of the cell body after a depolarizing event, how might this affect the ability of the neuron to respond to future excitatory signals?

43. Why does an "all-or-none" action potential in a presynaptic neuron not necessarily produce an "all-or-none" action potential in a postsynaptic neuron?

11.6   The PNS relays information between tissues and the CNS

11.7   The brain and spinal cord comprise the CNS

Crossword Puzzle

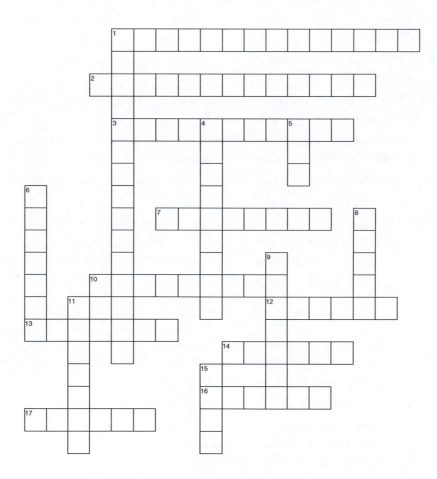

Across

1. The cell bodies of _____ neurons lie outside the CNS
2. The cell bodies of _____ neurons lie inside the CNS
3. The branch of the autonomic nervous system that prepares the body for "fight or flight"
7. Connective tissue membranes that surround the CNS
10. Cerebrospinal fluid is secreted into the _____ of the brain
12. Bundles of many axons wrapped in a connective sheath
13. _____ nerves that connect directly with the brain
14. The root of a spinal nerve that contains sensory neurons
16. A spinal _____ is an involuntary response that bypasses the brain
17. _____ nerves connect with the spinal cord

Down

1. The branch of the automatic nervous system that predominates during relaxation
4. The branch of the PNS that controls smooth and cardiac muscle
5. Pathways of the autonomic nervous system require _____ neurons
6. The branch of the PNS that controls skeletal muscle
8. The _____ matter of the CNS contains nerve tracts
9. The root of a spinal nerve that contains motor neurons
11. A collection of cell bodies of postsynaptic neurons
15. The _____ matter of the CNS contains unmyelinated axons and cell bodies

11.8   The brain processes and acts on information

11.9   Brain activity continues during sleep

11.10   The limbic system is the site of emotion and basic behaviors

11.11   Memory involves storing and retrieving information

Matching

Select from the following list the correct term for each definition below.

hypothalamus   midbrain   cerebrum   pons

corpus callosum   cerebellum   thalamus   hindbrain

medulla oblongata   forebrain

_____ 1. receives, processes, and transfers information

_____ 2. one of three main divisions of the brain, involved in emotions and conscious thought

_____ 3. coordinates basic subconscious body movements

_____ 4. plays an important role in regulating homeostasis

_____ 5. aids the medulla in regulating respiration

_____ 6. one of the three main divisions of the brain, functions in vision, hearing, and unconscious skeletal muscle movements

_____ 7. controls automatic functions of internal organs

_____ 8. bridges the two cerebral hemispheres

_____ 9. one of the three main divisions of the brain, includes the medulla oblongata, cerebellum, and pons

_____ 10. an area in the forebrain involved in decision making and conscious thought

Fill-in-the-Blank

All statements reference section 11.8.

11. The cardiovascular center is located in the _____ _____.

12. Motor nerves that leave the forebrain cross over to the other side of the body in the

    _____ _____.

13. Wakefulness is maintained by the _____ _____.

14. Four important areas of the brain located in the forebrain include the _____,

    _____, _____ _____, and

    _____.

15. Short-term memory is created in the _____ _____.

16. The _____ lobe of the brain initiates muscle activity and conscious thought.

Labeling/Coloring

Select different colors for each structure listed below. Color the coding circles and the corresponding structures in Figure 11.2.

○ pons   ○ cerebrum

○ cerebellum   ○ thalamus

○ medulla oblongata   ○ corpus callosum

Figure 11.2

11.12 Psychoactive drugs affect higher brain functions
11.13 Disorders of the nervous system

Short Answer

1. What is a psychoactive drug?

2. Why can psychoactive drugs affect the brain when many other drugs cannot?

3. What is the common mode of action for psychoactive drugs?

4. What is the effect of cocaine on the brain?

5. Why does tolerance develop?

6. Define *addiction*.

Completion

Name each disorder of the nervous system described in the table.

Disorder	Description
7.	blow to the head or neck with loss of consciousness
8.	violent trauma to the vertebrae that compresses, tears, or severs the spinal cord
9.	an infection resulting in inflammation of the brain
10.	an infection resulting in inflammation of the meninges
11.	viral brain infection transmitted to humans by direct contact with infected animals, often through a bite
12.	recurrent episodes of abnormal electrical activity in the brain, often of unknown causes
13.	progressive degenerative disorder caused by the loss of dopamine-releasing neurons
14.	changes in structure and activity of frontal and temporal lobes leading to mental impairment
15.	abnormal growth in or on the brain

Looking Back/Looking Forward

1. Consider the phospholipid structure of the plasma membrane presented in Chapter 2 and Chapter 3. Why does the myelin sheath act as an insulator, allowing ions to flow between segments rather than the length of the axon membrane?

2. Chapter 22 looks at the process of evolution, in which successful organisms survive and reproduce. Why might evolution have led to CNS neurons with limited or no ability to regenerate if injured?

Chapter Test

Multiple Choice

1. Which of the following does not apply to the peripheral nervous system?
   a.   components are found outside the brain and spinal cord
   b.   includes a sensory and motor division
   c.   the somatic division controls smooth muscle activity
   d.   the sympathetic and parasympathetic division work antagonistically

2. Neurons of the PNS that transmit information to the CNS are
   a.   sensory neurons.
   b.   motor neurons.
   c.   interneurons.
   d.   sympathetic neurons.

3. The _____ of a neuron carry(ies) information toward the cell body, while the _____ carry(ies) information away from the cell body.
   a.   myelin sheath, synaptic knob
   b.   dendrites, axon
   c.   axon, dendrites
   d.   postsynaptic membrane, presynaptic membrane

4. PNS neurons have the ability to regenerate if the axon is damaged, while CNS neurons do not. This is because
   a.   CNS neurons lack a myelin sheath.
   b.   the myelin sheath of a CNS neuron does not have the ability to regenerate.

    c.   the myelin sheath of CNS neurons degenerates if the neuron is damaged.

    d.   the cell body of CNS neurons prevents regeneration if the axon is damaged.

5. The sodium-potassium pump

    a.   transports $3Na^+$ ions out of the cell and $2K^+$ ions into the cell.

    b.   transports $3Na^+$ ions into the cell and $2K^+$ ions out of the cell.

    c.   helps maintain the resting membrane potential of a cell.

    d.   a and c

    e.   b and c

6. Which of the following characteristics of the neuron cell membrane contribute to the maintenance of the resting membrane potential?

    a.   The membrane is more permeable to sodium than to potassium.

    b.   Sodium leaks out of the membrane more easily than potassium.

    c.   The sodium-potassium pump transports three sodium ions out of the cell for every two potassium ions it pumps into the cell.

    d.   a and c

    e.   all of the above

7. Graded potentials

    a.   are permanent changes in the resting potential of a membrane.

    b.   occur in a local area of the membrane.

    c.   always depolarize the membrane.

    d.   grow stronger as they spread through the cell.

8. Summation may occur

    a.   when rapid stimulation of a neuron causes the membrane voltage to reach threshold levels.

    b.   when many neurons all cause hyperpolarizing graded potentials.

    c.   when the presence of the axon hillock lowers the normal threshold level.

    d.   when the cell body contains both chemically sensitive and voltage-sensitive channels.

9. Repolarization occurs when

    a.   voltage-sensitive calcium channels open.

    b.   chemically sensitive sodium channels close.

    c.   potassium channels open.

    d.   sodium leaks out of the cell.

10. Which of the following would occur in a neuron if calcium could not enter the axon bulb?

    a.   absence of graded potentials

    b.   absence of action potentials

    c.   inability to manufacture neurotransmitters

    d.   inability to release neurotransmitters

11. The absolute refractory period
    a. occurs during transmission of an action potential.
    b. occurs shortly after transmission of an action potential.
    c. occurs shortly before transmission of an action potential.
    d. reflects the actions of the sodium-potassium pump.

12. Action potentials are "all-or-none" events. Which of the following statements does not accurately reflect the "all or none" characteristic of an action potential?
    a. Besides generating an action potential, the only additional information a neuron can transmit is the frequency of stimulation.
    b. If a threshold level of stimulation is reached, the action potential will always be of the same form and strength.
    c. Summation in a postsynaptic neuron can occur when the frequency of stimulation increases.
    d. The level of neurotransmitter bound to the postsynaptic membrane affects the speed of transmission of the action potential.

13. Excitatory neurotransmitters
    a. depolarize the postsynaptic membrane.
    b. hyperpolarize the postsynaptic membrane.
    c. always initiate an action potential.
    d. can depolarize or hyperpolarize the postsynaptic membrane depending on the type of receptors present.

14. Which of the following must occur immediately before neurotransmitters can be released into the synaptic cleft?
    a. an action potential
    b. a graded potential
    c. reuptake of neurotransmitter by the presynaptic membrane
    d. opening of calcium channels on the presynaptic membrane

15. When one neuron receives input from many neurons, this is called
    a. summation.
    b. convergence.
    c. divergence.
    d. multiple stimulation.

16. The ventral root of a spinal nerve contains
    a. sensory neurons.
    b. motor neurons.
    c. sensory and motor neurons.
    d. cell bodies only.

17. Motor pathways requiring a preganglionic and a postganglionic neuron belong to the _____ nervous system.
    a. somatic division of the
    b. sensory division of the
    c. central
    d. autonomic

18. If your heart rate and blood pressure increase in response to a crisis, you are experiencing the immediate effects of the _____ nervous system.
    a. sensory
    b. sympathetic
    c. parasympathetic
    d. reflexive

19. The patellar reflex
    a. is a spinal reflex.
    b. occurs with little or no involvement of the brain.
    c. is voluntary.
    d. a and b.
    e. a, b, and c.

20. White matter of the spinal cord contains _____, while gray matter contains _____.
    a. nerve tracts, cell bodies
    b. cell bodies, nerves
    c. unmyelinated axons, myelinated axons
    d. motor neurons, sensory neurons

21. Which of the following major functional divisions of the brain is/are incorrectly matched with its function?
    a. *midbrain*—coordinates muscle groups
    b. *hindbrain*—responds to light and sound
    c. *forebrain*—integrates sensory information
    d. all of the above are correctly matched

22. Long-term memory occurs when
    a. neurons are permanently changed.
    b. the memory is stored in the cerebellum.
    c. dopamine remains in the synaptic cleft.
    d. convergence strengthens the excitatory effect of a neurotransmitter.

23. In the primary somatosensory area
    a. the brain responds to visual input.
    b. conscious thought is coordinated with motor responses.
    c. body parts that are extremely sensitive are represented by a greater area than less sensitive body parts.
    d. body parts that are coordinated with motor responses are represented by a greater area than body parts that are not.

24. During which stage of sleep do skeletal muscles relax and eye and body movements cease?

   a.   stage 1

   b.   stage 2

   c.   stage 3

   d.   stage 4

   e.   stage 5

25. The site of emotions and basic behavior is the

   a.   parietal lobe of the cerebral cortex.

   b.   thalamus.

   c.   limbic system.

   d.   hindbrain.

Self Test

Each of the Key Concepts listed at the beginning of this chapter has been rewritten as a question below. After successfully completing the study guide exercises and the Chapter Test, you should be able to answer each of these questions. Refer to the Key Concepts list at the beginning of this chapter to check your answers.

1. What are the three characteristics of the nervous system?

2. What are the two principal parts of the nervous system and what components does each contain?

3. What cells of the nervous system are specialized to receive and transmit information?

4. What are the three classifications of neurons?

5 What are the parts of a neuron?

6. What is the resting membrane potential?

7. What causes a graded potential to occur?

8. What is an action potential?

9. What are the three events of an action potential?

10. During an action potential, what ion activity causes each of the three events?

11. What is a neurotransmitter? How do excitatory neurotransmitters differ from inhibitory neurotransmitters in their effects on the postsynaptic cell membrane?

12. What synaptic cells integrate and process information?

13. What is a nerve? How do spinal nerves differ from cranial nerves?

14. What are the two divisions of the peripheral nervous system? Which division carries information to the brain? Which division carries information away from the brain?

15. What are the branches of the motor nervous system? What are the branches of the autonomic division?

16. What are the regions of the brain?

17. What is the general function of the cerebrum? List the lobes of the cerebrum.

18. What is the function of the reticular activating system? What is the function of the limbic system?

19. How does short-term memory differ from long-term memory in location and duration?

20. How do psychoactive drugs affect the brain?

21. What are the four general categories of nervous system disorders discussed in this chapter?

12

Sensory Mechanisms

Chapter Summary and Key Concepts

After reading and studying this chapter, you should know the following:

Sections 12.1, 12.2, 12.3

1. Receptors convert the energy of a stimulus into impulses in sensory neurons.

2. Receptors are classified according to the type of stimulus energy they convert.

3. The brain interprets nerve impulses based on their origin and frequency.

4. Adaptation occurs in some receptors and is the cessation of impulses in the presence of a continued stimulus.

5. The somatic senses are temperature, touch, vibration, pressure, pain, and awareness of body position and movements. The somatic senses arise from receptors located throughout the body.

6. The special senses are taste, smell, hearing, balance, and vision. The special senses arise from receptors in specialized areas of the body.

7. Taste depends on chemoreceptors located in taste buds on the tongue and mouth.

8. Smell depends on chemoreceptors located in the nasal passages.

Sections 12.4, 12.5

9. Mechanoreceptors in the ear detect sound waves.

10. The ear consists of an outer ear, middle ear, and inner ear.

11. The outer ear is called the pinna and channels sound waves.

12. The middle ear consists of three small bones that amplify sound and transmit the waves to the inner ear.

13. The inner ear converts sound to impulses in the cochlea and transmits information to the brain via the auditory nerve.

14. The inner ear also contains the semicircular canals that sense rotational movement, and the vestibular apparatus that senses static head position and linear acceleration.

Sections 12.6, 12.7

15. The structures of the eye convert visual stimuli to impulses that are transmitted to the brain via the optic nerve.

16. Light entering the eye is bent by the cornea and the lens to focus an image.

17. Accommodation is the adjustment of lens curvature that allows the eye to focus on objects at any distance.

18. The photoreceptor cells of the eye are the photopigment-containing rods and cones.

19. Rods function in dim light, and cones respond to color.

20. Disorders of sensory mechanisms include retinal detachment, glaucoma, cataracts, color blindness, otitis media, and Ménière's syndrome.

Exercises

Complete the exercises for each section after you have read and studied the section. If you cannot answer some questions, or if you answer them incorrectly, return to the chapter and review this information again. You may find it helpful to work on only one section at a time. When you have completed all sections, take the Chapter Test as an indicator of your mastery of this topic.

12.1 Receptors receive and convert stimuli

12.2 Somatic sensations arise from receptors throughout the body

12.3 Taste and smell depend on chemoreceptors

Define

Write the term(s) for each definition.

_____ 1. sensory input that causes some change within or outside the body

_____ 2. a structure specialized to receive certain stimuli

_____ 3. mechanoreceptors that monitor muscle length

_____ 4. most common receptor type for fast pain

_____ 5. most common receptor type for slow pain

_____ 6. chemicals that simulate hairs on taste cells

_____ 7. recently discovered fifth taste

_____ 8. receptor cells that detect odors

Completion

Write the stimulus for each skin mechanoreceptor in the table.

Mechanoreceptor	Stimulus
9. Unencapsulated dendritic endings	
10. Merkel discs	
11. Pacinian corpuscles	
12. Meissner's corpuscles	
13. Ruffini endings	

Short Answer

14. How does a sensation differ from perception?

15. List the five types of receptors and the type of stimulus to which each responds.

16. What are "silent" receptors and what is their function?

17. List the six sensations classified as somatic sensations.

18. What specific part of the brain receives information about somatic sensations, and what part of the brain initiates a response?

19. What are the four primary tastes and where are receptors for each located on the tongue?

20. Describe two differences between taste and smell.

Critical Thinking

21. Sensory adaptation normally confers a survival advantage. Why can olfactory adaptation sometimes be dangerous? Why does adaptation of pain receptors not confer a survival advantage?

12.4   Hearing: Mechanoreceptors detect sound waves

12.5   The inner ear plays an essential role in balance

Labeling

Use the terms to label Figure 12.1.

middle ear	**pinna**	**malleus**
auditory canal	**semicircular canals**	**oval window**
outer ear	**cochlear nerve**	**vestibular nerve**
incus	**cochlea**	**tympanic membrane**
vestibule	**inner ear**	**stapes**
round window	**auditory tube**	

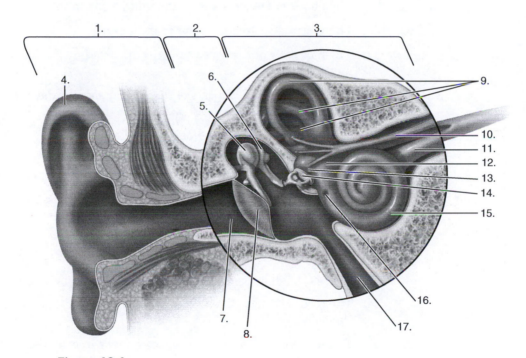

Figure 12.1

1. _____
2. _____
3. _____
4. _____
5. _____
6. _____
7. _____
8. _____
9. _____

10. _____
11. _____
12. _____
13. _____
14. _____
15. _____
16. _____
17. _____

True/False

Mark each statement as true or false. If you mark a statement as false, replace the underlined term or phrase with the term or phrase needed to make the statement true.

_____ 18. Sounds are waves of compressed <u>air</u>.

_____ 19. The loudness of a sound depends on the <u>amplitude</u> of the sound waves.

_____ 20. The tone of a sound depends on the <u>number</u> of the waves.

_____ 21. Sound waves arrive first at the <u>tympanic membrane</u>.

_____ 22. The <u>middle</u> ear consists of the pinna and the auditory canal.

_____ 23. Vibrations of the <u>pinna</u> cause the malleus, incus, and stapes to vibrate.

_____ 24. Vibrations of the stapes cause the <u>round window</u> to vibrate.

_____ 25. The auditory tube runs from the middle ear to the <u>inner ear</u>.

_____ 26. The <u>vestibule</u> converts sound to impulses.

_____ 27. The hair cells of the inner ear are supported by the <u>basilar membrane</u>.

_____ 28. The organ of Corti consists of the <u>basilar membrane</u> and the tectorial membrane.

_____ 29. The hair cells of the inner ear bend in response to vibrations of the <u>tectorial membrane</u>.

_____ 30. Impulses generated by hair cells in the inner ear are carried first by the <u>vestibulocochlear</u> nerve.

_____ 31. High-pitched tones vibrate the basilar membrane close to the <u>tip</u> of the cochlea.

_____ 32. The <u>middle ear</u> amplifies sound before passing it to the cochlea.

_____ 33. The <u>semicircular canals</u> sense rotational movement of the head.

_____ 34. The <u>cochlea</u> senses linear acceleration and deceleration of the body.

_____ 35. The <u>utricle and saccule</u> can sense changes in gravitational pull.

12.6 Vision: Detecting and interpreting visual stimuli

12.7 Disorders of sensory mechanisms

Labeling

Figure 12.2

1. _____	9. _____
2. _____	10. _____
3. _____	11. _____
4. _____	12. _____
5. _____	13. _____
6. _____	14. _____
7. _____	15. _____
8. _____	

Matching

Match each of the following structures of the eye with its correct function.

_____ 16. **sclera**   a. absorbs light and converts it into impulses

_____ 17. **cornea**   b. central region of retina with the highest density of photoreceptors

_____ 18. **aqueous humor**   c. nourishes and cushions cornea and lens

_____ 19. **iris**   d. regulates focus

_____ 20. **lens**   e. covers and protects eyeball

_____ 21. **ciliary muscle**   f. "blind spot" where optic nerve exits eye

_____ 22. **vitreous humor**   g. transmits light to retina

_____ 23. **retina**   h. nourishes retina and absorbs light not absorbed by retinal photoreceptors

_____ 24. **retinal rods**   i. photoreceptors responsible for black-and-white vision in dim light

_____ 25. **retinal cones**   j. transmits impulses to the brain

_____ 26. **macula**   k. bends incoming light and focuses light

_____ 27. **optic disk**   l. photoreceptors responsible for color vision and high visual acuity

_____ 28. **optic nerve**   m. adjusts amount of incoming light

_____ 29. **choroid**   n. adjusts curvature of the lens

Fill-in-the-Blank

Referenced sections are in parentheses.

30. The flattened disks of photoreceptor cells contain a protein called a _____. (12.6)

31. _____ is the photopigment found in rods. (12.6)

32. _____ are photoreceptor cells that respond to color. (12.6)

33. Light entering the eye is bent first by the _____ and then by the _____. (12.6)

34. The innermost layer of the retina consists of neurons called _____ _____. (12.6)

35. Nearsightedness, also called _____, occurs when the eyeball is too _____. (12.6)

Completion

Write in the table the name of each sensory disorder described.

Sensory Disorder	Description
36.	the retina separates from the choroid; often caused by a blow to the head
37.	blockage in the Canal of Schlemm prevents drainage of the aqueous humor, leading to pressure in the eye
38.	a decrease in the normal transparency of the lens
39.	inflammation of the middle ear
40.	a chronic condition of the inner ear, characterized by recurrent episodes of dizziness, nausea, and hearing loss
41.	distant objects are focused in front of the retina

Looking Back/Looking Forward

1. Chapter 11 discussed the nervous system and the role of neurotransmitters. Endorphins bind to opiate receptors, the same receptors that bind morphine. Why can activities that increase the level of circulating endorphins, such as stress and acupuncture, help to reduce pain?

2. Chapter 14 presents the digestive system. Amylase is an enzyme found in saliva that breaks down complex carbohydrates into simple sugars. The elderly often experience a decrease in salivary sensations. How does this affect their sense of taste?

Chapter Test

Multiple Choice

1. Receptors are classified according to
   a. their location in the body.
   b. the type of cell they arise from.
   c. the type of stimulus energy they convert.
   d. how easily they generate an action potential.

2. Pain receptors adapt slowly or not at all because
   a. lack of adaptation is important to survival.
   b. only mechanoreceptors are capable of adaptation.
   c. adaptation occurs but is ignored by the brain.
   d. the stimulus is too strong for adaptation to occur.

3. Pacinian corpuscles detect
   a. light touch.
   b. deep pressure.
   c. temperature.
   d. ongoing pressure.

4. Muscle spindles are
   a. chemoreceptors.
   b. photoreceptors.
   c. thermoreceptors.
   d. mechanoreceptors.

5. Somatic sensations are carried by sensory neurons from receptors to the
   _____ of the brain.
   a. parietal lobe of the cerebral cortex
   b. frontal lobe of the cerebral cortex
   c. cerebellum
   d. medulla oblongata

6. Chemoreceptors for taste are specific for a chemical called a
   a. tastant.
   b. papilla.
   c. taste bud.
   d. taste hair.

7. The chemoreceptors for smell are located on
   a. sensory neurons.
   b. epithelial cells.
   c. naked nerve endings.
   d. glandular cells.

8. The olfactory bulb is
    a. the site of origin for olfactory receptor cells.
    b. an area of the nose where olfactory cells are highly concentrated.
    c. an area of the brain where olfactory receptor cells synapse with other neurons.
    d. the area of the brain where smells are interpreted.

9. Sounds are waves of compressed air. The loudness of a sound is determined by
    a. the frequency of the waves.
    b. the amplitude of the waves.
    c. the temperature of the air.
    d. the volume of air.

10. The pinna and auditory canal are part of the
    a. inner ear.
    b. middle ear.
    c. outer ear.
    d. temporal lobe.

11. Which part of the ear generates impulses for sound?
    a. ampulla
    b. utricle
    c. saccule
    d. cochlea

12. Vibrations are passed from the middle ear to the inner ear through the
    a. cochlear window.
    b. round window.
    c. oval window.
    d. tympanic window.

13. The base of the cochlear duct is formed by the
    a. tectorial membrane.
    b. vestibular membrane.
    c. cochlear membrane.
    d. basilar membrane.

14. The structure of the inner ear that converts pressure waves to impulses is the
    a. vestibular canal.
    b. cochlear duct.
    c. organ of Corti.
    d. round window.

15. The mechanoreceptor cells of the ear are
    a. hair cells.
    b. tectorial cells.
    c. vestibular cells.
    d. tympanic cells.

16. The semicircular canals sense
    a. rotational movement of the head and body.
    b. inertia.
    c. linear acceleration.
    d. vertical deceleration.

17. Which of the following structural layers of the eye is *incorrectly* matched with its function?
    a. sclera—produces the aqueous humor
    b. retina—contains photoreceptor cells
    c. choroids—contains blood vessels that nourish the retina
    d. all are correctly matched with their function

18. The photoreceptors of the eye are
    a. the optic nerves.
    b. the ciliary muscles.
    c. rods and cones.
    d. choroid pigment cells.

19. What structure of the eye determines how much light enters the eye?
    a. cornea
    b. pupil
    c. iris
    d. lens

20. The fluid found between the lens and the cornea is the
    a. aqueous humor.
    b. vitreous humor.
    c. ciliary fluid.
    d. synovial fluid.

21. Accommodation occurs as a result of the actions of the
    a. cornea.
    b. ciliary muscle.
    c. vitreous humor.
    d. retina.

22. The layer of bipolar cells in the retina
    a. receives information from the rods and cones.
    b. receives information from ganglion cells.
    c. absorbs light to stimulate rods and cones.
    d. possesses axons that will become the optic nerve.

23. Rhodopsin is
    a.   an enzyme produced by the ciliary muscle.
    b.   a nutrient found in the vitreous humor.
    c.   the photopigment found in rods.
    d.   a chemoreceptor that monitors blood supply to the retina.

24. The optic disk is the area of the eye
    a.   with the highest concentration of cones.
    b.   with the highest concentration of rods.
    c.   where the optic nerve exits.
    d.   where the retina attaches to the sclera.

25. Ménière's syndrome is a condition in which
    a.   the retina detaches from the choroids.
    b.   an individual suffers from dizziness and hearing loss.
    c.   the middle ear is inflamed.
    d.   olfactory receptors are damaged.

Self Test

Each of the Key Concepts listed at the beginning of this chapter has been rewritten as a question below. After successfully completing the study guide exercises and the Chapter Test, you should be able to answer each of these questions. Refer to the Key Concepts list at the beginning of this chapter to check your answers.

1. What do receptors do with a stimulus?

2. How are receptors classified?

3. What characteristics does the brain use to interpret a nerve impulse?

4. What is adaptation? Does it occur in all receptors?

5. What are the somatic senses? Where are the receptors for somatic senses located?

6. What are the special senses? Where are the receptors for the special senses located?

7. What type of receptor is responsible for taste? Where are these receptors located?

8. What type of receptor is responsible for smell? Where are these receptors located?

9. What type of receptors detect sound waves?

10. What are the three main parts of the ear?

11. What is the outer ear called? What is the function of the outer ear?

12. What structures are found in the middle ear? What is the function of the middle ear?

13. What role does the inner ear play in sound? What general part of the inner ear is responsible for hearing?

14. What part of the inner ear functions in sensing rotational movement? What part of the inner ear functions in sensing head position and linear acceleration?

15. What does the eye do with visual stimuli?

16. What structures of the eye bend light to focus an image?

17. What is accommodation?

18. What are the photoreceptor cells of the eye?

19. What is the specific function of rods and cones?

20. What are the six disorders of sensory mechanisms discussed in this chapter?

13

The Endocrine System

Chapter Summary and Key Concepts

After reading and studying this chapter, you should know the following:

Sections 13.1, 13.2, 13.3

1. The endocrine system is a collection of specialized cells, tissues, and glands that produce hormones. Hormones are secreted into extracellular space and enter the blood or lymph.

2. Hormones act only on their target cells.

3. Endocrine control of body activities is slower than nervous control.

4. Hormones are classified as steroid or nonsteroid.

5. Steroid hormones are synthesized from cholesterol, are lipid soluble, and exert their effect by binding to intracellular receptors and activating or inhibiting genes.

6. Nonsteroid hormones are synthesized from amino acids, are not lipid soluble, and exert their effect by binding to a cell membrane and activating a second messenger molecule.

7. Many hormones participate in negative feedback loops that maintain homeostasis.

8. The hypothalamus is a region of the brain that interacts with the endocrine system primarily through the pituitary gland.

9. The pituitary gland consists of a posterior lobe and an anterior lobe.

10. The posterior pituitary stores ADH and oxytocin secreted by neuroendocrine cells in the hypothalamus.

11. The anterior pituitary produces ACTH, TSH, FSH, LH, PRL, and GH. Releasing and inhibiting hormones produced by the hypothalamus control the release of these hormones.

Sections 13.4, 13.5, 13.6, 13.7, 13.8

12. The pancreas secretes insulin when blood sugar levels are high, and glucagon when blood sugar levels are low. The pancreas also secretes somatostatin, a hormone that inhibits insulin and glucagon secretion.

13. The adrenal cortex produces cortisol that helps maintain blood glucose levels, and aldosterone that helps maintain water balance. The adrenal cortex also produces small amounts of sex hormones.

14. The adrenal medulla produces epinephrine and norepinephrine. The release of these hormones is stimulated by the sympathetic nervous system, and they contribute to the fight-or-flight response.

15. The thyroid gland produces thyroxin, which regulates the production of ATP, and calcitonin, which decreases blood calcium levels.

16. The parathyroid glands produce parathyroid hormone, which increases the blood calcium levels.

17. The testes produce testosterone and the ovaries produce estrogen and progesterone. These sex hormones regulate sexual development, activities, and secondary sex characteristics.

18. The thymus gland secretes thymosin and thymopoietin for the maturation of T cells; the pineal gland secretes melatonin, which may play a role in sleep/wake cycles; and the heart, kidneys, and digestive system all secrete hormones specialized for their functions.

Sections 13.9, 13.10

19. Histamine, prostaglandins, nitric oxide, and growth factors are chemical messengers similar to hormones but with only local effects.

20. Disorders of the endocrine system include diabetes mellitus, hyperthyroidism, hypothyroidism, Addison's disease, and Cushing's syndrome.

Exercises

Complete the exercises for each section after you have read and studied the section. If you cannot answer some questions, or if you answer them incorrectly, return to the chapter and review this information. You may find it helpful to work on only one section at a time. When you have completed all sections, take the Chapter Test as an indicator of your mastery of this topic.

13.1   The endocrine system produces hormones

13.2   Hormones are classified as steroid or nonsteroid

13.3   The hypothalamus and the pituitary gland

Define

Write the term(s) for each definition.

_____ 2. bloodborne units of information secreted by glands

_____ 3. groups of cells that secrete hormones into the extracellular space

_____ 4. cells responsive to a particular hormone

_____ 5. lipid-soluble hormones synthesized from cholesterol

_____ 6. mostly lipid-insoluble hormones synthesized from amino acids

_____ 7. a communication system that uses chemical messengers

_____ 8. a small region in the brain that serves as a homeostatic control center

_____ 9. a small endocrine gland called the "master gland"

_____ 10. cells in the hypothalamus that function as both nerve cells and endocrine cells

_____ 7. an inactive molecule in a cell converted to an active molecule by the binding of a hormone to the cell membrane

Fill-in-the-Blank

Referenced sections are in parentheses.

11. Hormones circulate in the _____ and act only on specific _____. (13.1)

12. There are approximately _____ known hormones. (13.1)

13. The target cells of a hormone will have the right _____ for that particular hormone. (13.1)

14. The _____ system and the _____ system interact with each other to maintain homeostasis. (13.1)

15. _____ hormones enter the cell to carry out their effects. (13.2)

16. _____ hormones may utilize a second messenger to carry out their effects. (13.2)

17. A hormone-receptor complex is involved in the action of _____ hormones. (13.2)

18. A second messenger is involved in the action of _____ hormones. (13.2)

19. In a negative feedback loop involving a hormone, the effectors are the _____

_____ _____. (13.2)

20. In addition to producing hormones, the hypothalamus monitors the _____

_____, _____ _____,

_____, and _____ _____. (13.3)

21. Neuroendocrine cells have cell bodies located in the _____ and axon endings

located in the _____ _____. (13.3)

22. The connection between the hypothalamus and the posterior pituitary is

_____, while the connection between the hypothalamus and the anterior

pituitary is _____. (13.3)

23. Diabetes insipidus may be caused by inadequate amounts of _____. (13.3)

Labeling/Coloring

Select different colors for each structure listed below. Color the coding circles and the corresponding structures in Figure 13.1.

○ anterior lobe of the pituitary

○ adrenal gland

○ pituitary gland

○ posterior lobe of the pituitary

○ hypothalamus

○ parathyroid gland

○ kidney

○ thyroid gland

○ thymus gland

○ stomach

○ testis

○ heart

○ intestines

○ ovary

○ pancreas

○ pineal gland

Figure 13.1

Completion

For each hormone listed, select the correct action and target cell.

Action	Target Cell
a. stimulates synthesis and release of thyroid hormone	i. kidney
b. stimulates ovulation and estrogen secretion in females, and testosterone secretion in males	j. adrenal cortex
c. reduces water lost in urine	k. uterus
d. stimulates glucocorticoid release	l. mammary glands
e. stimulates contractions in the uterus and milk ejection from the mammary glands	m. ovaries and testes
f. stimulates growth in children and metabolism in adults	n. most cells
g. stimulates mammary gland development and milk production	o. ovaries and testes
h. stimulates egg maturation and estrogen secretion in females and sperm production in males	p. thyroid gland

_____   24. ADH

_____   25. Oxytocin

_____   26. Adrenocorticotropic hormone

_____   27. Thyroid-stimulating hormone

_____   28. Follicle-stimulating hormone

_____   29. Leutinizing hormone

_____   30. Prolactin

_____   31. Growth hormone

Short Answer

32. Both the endocrine system and the nervous system send messages that affect cells. List four characteristics that describe the operation of the endocrine system.

33. Compare and contrast steroid and nonsteroid hormones by describing:

    a. which chemical molecule is utilized in hormone synthesis

    b. lipid solubility

    c. how the hormones exert their effect on a target cell

34. Why are steroid hormones usually slower acting than nonsteroid hormones?

Critical Thinking

35. Both the nervous system and the endocrine system carry information to the body. How is the endocrine system different from the nervous system in its delivery of messages?

13.4 The pancreas secretes glucagon, insulin, and somatostatin

13.5 The adrenal gland comprises the cortex and medulla

13.6 Thyroid and parathyroid glands

13.7 The testes and ovaries produce sex hormones

13.8 Other glands and organs also secrete hormones

Completion

For each hormone listed in the table, fill in the site of production, action, and target cell(s).

Hormone	Site of Production	Action	Target Cell
1. glucagon			
2. insulin			
3. somatostatin			
4. cortisol			
5. aldosterone			

Hormone	Site of Production	Action	Target Cell
6. epinephrine			
7. thyroxine			
8. parathyroid hormone			
9. testosterone			
10. estrogen			
11. thymosin			
12. melatonin			
13. cholecystokinin			
14. erythropoietin			

Crossword Puzzle

Write in the hormone that will be secreted in response to each physiological situation represented in the clue.

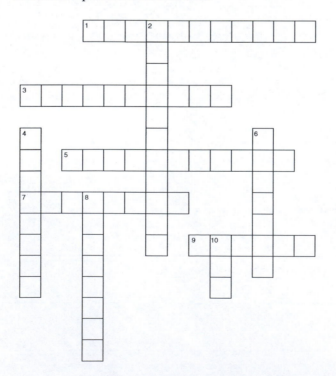

Across

1. _____ hormone is secreted when blood calcium levels are too low
3. Blood calcium levels are too high
5. A sudden crisis activates the sympathetic nervous system
7. Injury and emotional stress
9. _____ natriuretic hormone is secreted when blood pressure rises

Down

2. Potassium levels in the body are too high
4. Blood sugar levels are too low
6. Blood glucose levels increase
8. A situation that creates an increased need for energy
10. Iodine deficiency leads to absence of thyroxin

Critical Thinking

18. A patient currently taking insulin for diabetes is diagnosed with lupus erythematosus, an autoimmune disorder. The doctor prescribes steroids to combat the inflammation caused by lupus. Why would the prescription of these cortisol-like drugs present a problem for the patient?

13.9   Other chemical messengers

13.10   Disorders of the endocrine system

Completion

Complete the table below, writing in the action associated with each chemical messenger.

Chemical Messenger	Primary Action
Histamine	1.
Prostaglandins	2.
Nitric oxide	3.
Growth factor	4.

For each of the endocrine disorders listed below, write in the cause and symptoms.

5. Hypothyroidism

   a.   cause:

   b.   symptoms:

6. Hyperthyroidism
   a. cause:

   b. symptoms:

7. Addison's disease
   a. cause:

   b. symptoms:

8. Cushing's syndrome
   a. cause:

   b. symptoms:

9. Type I diabetes
   a. cause:

   b. symptoms:

10. Type II diabetes
    a. cause:

    b. symptoms:

Looking Back/Looking Forward

1. Chapter 4 described positive and negative feedback control systems. Positive feedback control is rarely seen in the endocrine system. Why do you think negative feedback control is the predominant form of regulation in the endocrine system?

2. Chapter 15 describes how anti-diuretic hormone (ADH) affects urine production by the kidneys. Excessive alcohol consumption can lead to a significant reduction in ADH production, as well as vomiting, and can be fatal. What potentially lethal condition can be caused by excessive alcohol consumption?

Chapter Test

Multiple Choice

1. Hormones secreted into interstitial fluid, lymph, and blood will have been produced by
   a. ductless glands
   b. endocrine glands
   c. exocrine glands
   d. a and b
   e. a and c

2. The target cells of an endocrine gland
   a. have specific receptors for the hormone.
   b. respond to all endocrine hormones released by all endocrine glands.
   c. will always be in close proximity to the gland.
   d. are connected to the gland by nerve cells.

3. Which of the following statements is not an accurate description of the roles of the nervous and endocrine systems in regulating body processes?
   a. the nervous and endocrine systems often interact with each other
   b. regulation by the endocrine system requires more time than regulation by the nervous system
   c. endocrine regulation is most effective for processes requiring long-term controls
   d. most reflexes depend on endocrine control

4. The nervous system and the endocrine system interact through the
   a. thyroid gland and the pituitary gland.
   b. pituitary gland and the hypothalamus.
   c. hypothalamus and the thyroid gland.
   d. anterior pituitary and posterior pituitary.

5. Steroid hormones exert their effect using
   a. a second messenger.
   b. cyclic AMP.
   c. a hormone-receptor complex.
   d. a carbohydrate molecule.

6. Nonsteroid hormones
   a. cross the cell membrane easily.
   b. bind to receptors on the cell surface.
   c. cross the membrane or bind to receptors, depending on the target cell.
   d. pass through protein channels in the membrane.

7. Nonsteroid hormones are usually faster acting than steroid hormones because
   a. they are transported more quickly.
   b. they activate proteins that already exist within the cell.
   c. they utilize enzymes.
   d. they are not attacked by immune cells.

8. The hypothalamus is able to produce hormones because
   a. it is in the brain.
   b. it contains neurosecretory cells.
   c. it is influenced by the pituitary gland.
   d. it is connected to the thalamus, which functions as a gland.

9. All of the following are accurate descriptions of the hypothalamus *except*
   a. neuroendocrine cells in the hypothalamus store hormones in the posterior pituitary.
   b. ADH stimulates the hypothalamus to produce oxytocin.
   c. the hypothalamus contains the cell bodies of neuroendocrine cells.
   d. the hypothalamus functions in both the nervous system and the endocrine system.

10. Which of the following hormones is *not* produced by the anterior pituitary?
    a. adrenocorticotropic hormone
    b. luteinizing hormone
    c. prolactin
    d. antidiuretic hormone

11. Hormones produced in the anterior pituitary will be released in response to _____ produced by the _____.
    a. factors, thyroid
    b. releasing hormones, anterior pituitary
    c. releasing hormones, posterior pituitary
    d. releasing hormones, hypothalamus

12. FSH stimulates
    a. the release of growth hormone by the pituitary gland.
    b. egg development in females and sperm development in males.
    c. ovulation in females and testosterone production in males.
    d. uterine contraction and milk production in pregnant or lactating females.

13. Diabetes insipidus
    a. results in decreased production of insulin.
    b. is caused by a virus.
    c. is characterized by excessive water loss through the kidneys.
    d. is caused by lack of ADH or poor response to ADH.
    e. a and c
    f. c and d

14. The _____ is both an endocrine and an exocrine gland.
    a. thyroid gland
    b. pituitary gland
    c. hypothalamus
    d. pancreas

15. Which of the following hormones will be most active after a 24-hour fast?
    a. insulin
    b. aldosterone
    c. glucagon
    d. somatostatin

16. Cortisol production occurs in the _____ and is stimulated by _____.
    a. adrenal medulla, FSH
    b. adrenal cortex, ACTH
    c. thyroid gland, TSH
    d. thymus gland, growth hormone

17. The adrenal medulla is similar to the hypothalamus because
    a. they are both neuroendocrine organs.
    b. they are both divided into two regions.
    c. they both produce inhibiting and releasing hormones.
    d. they both respond to pituitary hormones.

18. The blood calcium regulatory function of which gland is most important in adulthood?
    a. thyroid gland
    b. parathyroid gland
    c. pancreas
    d. thymus gland

19. Goiter is caused by
    a. increased levels of thyroxin.
    b. iodine deficiency.
    c. calcium deficiency.
    d. malfunction of the hypothalamus.

20. Which of the following is *not* a characteristic of parathyroid hormone?
    a. It increases blood calcium levels.
    b. It inhibits calcium retention by the kidneys.
    c. It increases calcium reabsorption by the digestive tract.
    d. It removes calcium from the bone.

21. Testosterone
    a. is produced by the anterior pituitary.
    b. is produced at consistent levels throughout life.
    c. secretion is stimulated by LH.
    d. inhibits the production of aldosterone.

22. The kidneys secrete
    a. secretin and cholecystokinin.
    b. renin and erythropoietin.
    c. histamine and prostaglandins.
    d. aldosterone and ADH.

23. Nitric oxide is not considered a hormone because
    a. it is not a steroid.
    b. it is short-lived.
    c. it is produced by epithelial tissue.
    d. it is not secreted into the bloodstream.

24. Which of the following sets of hormones oppose each other?
    a. secretin and renin
    b. FSH and LH
    c. glucagon and insulin
    d. progesterone and estrogen

25. The symptoms of Cushing's disease are caused by
    a. lowered levels of cortisol and aldosterone.
    b. hypersecretion by the thyroid gland.
    c. excessive levels of cortisol.
    d. excessive levels of testosterone.

Self Test

Each of the Key Concepts listed at the beginning of this chapter has been rewritten as a question below. After successfully completing the study guide exercises and the Chapter Test, you should be able to answer each of these questions. Refer to the Key Concepts list at the beginning of this chapter to check your answers.

1. What structures are included in the endocrine system? What is the initial destination of secreted hormones?

2. On what cells do hormones act?

3. How does endocrine control of body activities compare to nervous control in terms of speed?

4. What are the two general classifications of hormones?

5. What are the characteristics of steroid hormones?

6. What are the characteristics of nonsteroid hormones?

7. What type of feedback loops do hormones participate in?

8. What gland allows interaction between the hypothalamus and the endocrine system?

9. What are the two regions of the pituitary gland?

10. What hormones are stored in the posterior pituitary? Where were these hormones produced?

11. What are the hormones produced by the anterior pituitary? What controls the release of these hormones?

12. What are the three hormones produced by the pancreas? What is the function of each hormone?

13. What hormones are produced by the adrenal cortex? What are the functions of these hormones?

14. What hormones are produced by the adrenal medulla? What stimulates release of these hormones?

15. What hormones are produced by the thyroid gland? What are the functions of these hormones?

16. What hormone is produced by the parathyroid gland? What is the function of this hormone?

17. What hormones are produced by the testes and the ovaries? What are the functions of these hormones?

18. What are the names and functions of the hormones produced by the thymus gland and pineal gland?

19. What are the names of four chemical messengers?

20. What are at least four disorders of the endocrine system discussed in this chapter?

14

The Digestive System and Nutrition

Chapter Summary and Key Concepts

After reading and studying this chapter, you should know the following:

Sections 14.1, 14.2, 14.3, 14.4, 14.5

1. The digestive system provides nutrients to the body.

2. The walls of the gastrointestinal tract consist of four layers: the innermost mucosa, the submucosa, the muscularis, and the serosa.

3. Sphincters are thick rings of ciliary muscle that can close off the passageways between organs.

4. The five basic processes of the digestive system are mechanical processing and movement, secretion, digestion, absorption, and elimination.

5. Two types of motility important to the digestive system are peristalsis, which moves food forward, and segmentation, which mixes food.

6. In the mouth, the teeth grind the food and saliva begins the process of digestion.

7. The pharynx receives food from the mouth and initiates swallowing.

8. The esophagus is a muscular tube that connects the pharynx and the stomach.

9. The stomach functions to store and digest food, and to regulate the delivery of food to the small intestine.

10. The small intestine functions to digest carbohydrates, lipids, and proteins, and to absorb most of the nutrients.

Sections 14.6, 14.7, 14.8, 14.9

11. The accessory organs of the digestive system include the salivary glands, the pancreas, the liver, and the gallbladder.

12. The pancreas secretes digestive enzymes that complete digestion in the small intestine, and sodium bicarbonate that neutralizes stomach acid.

13. The liver produces bile, stores fat-soluble vitamins, stores glucose as glycogen, manufactures plasma proteins, synthesizes and stores some lipids, inactivates chemicals, converts ammonia to urea, and destroys worn-out red blood cells.

14. The gallbladder stores and releases bile.

15. The large intestine reabsorbs water and concentrates feces.

16. Proteins and carbohydrates are absorbed from the small intestine by active transport, lipids move by diffusion, and water moves by osmosis.

17. Digestion is regulated by the endocrine system and the nervous system.

Sections 14.10, 14.11, 14.12, 14.13

18. Carbohydrates, lipids, and proteins are essential nutrients in the diet.

19. Body weight depends on the balance between energy consumed and energy used.

20. Disorders of the digestive system include lactose intolerance, diverticulosis, colon polyps, hepatitis, gallstones, malnutrition, anorexia nervosa, and bulimia.

Exercises

Complete the exercises for each section after you have read and studied the section. If you cannot answer some questions, or if you answer them incorrectly, return to the chapter and review this information. You may find it helpful to work on only one section at a time. When you have completed all sections, take the Chapter Test as an indicator of your mastery of this topic.

14.1 The digestive system brings nutrients into the body

14.2 The mouth processes food for swallowing

14.3 The pharynx and esophagus deliver food to the stomach

14.4 The stomach stores, digests protein, and regulates delivery

14.5 The small intestine digests food and absorbs nutrients

Define

Write the term(s) for each definition.

_____ 1. the system that includes all organs responsible for getting nutrients into the body

_____ 2. a type of G.I. tract motility that propels food forward

_____ 3. a type of G.I. tract motility that mixes food

_____ 4. the visible region of a tooth

_____ 5. a structure that closes the airway to the trachea during swallowing

_____ 6. prevents reflux of the stomach's contents back into the esophagus

_____ 7. a mixture of partially digested food and gastric juice that enters the small intestine

_____ 8. the first region of the small intestine

_____ 9. the two regions of the small intestine where most absorption occurs

Labeling/Coloring

Select different colors for each structure listed below. Color the coding circles and the corresponding structures in Figure 14.1. Match the functions listed below with the correct structure by writing the letter of the function on the line next to each structure.

○ stomach   ○ pancreas

○ mouth   ○ liver

○ rectum   ○ pharynx

○ salivary glands   ○ esophagus

○ small intestine   ○ anus

○ large intestine   ○ gallbladder

○ sigmoid colon   ○ appendix

a. produces bile

b. terminal opening for expelling feces

c. swallowing is initiated here

d. secretes enzymes to digest all food groups

e. food is chewed and moistened here

f. secretes saliva

g. muscular tube that moves food from pharynx to stomach

h. reabsorbs water and concentrates undigested food into feces

i. passageway for fecal material

j. food mixes with gastric juices and protein digestion begins

k. organ where digestion of all food is completed

l. stores and concentrates bile

m. no known digestive function

n. stores feces

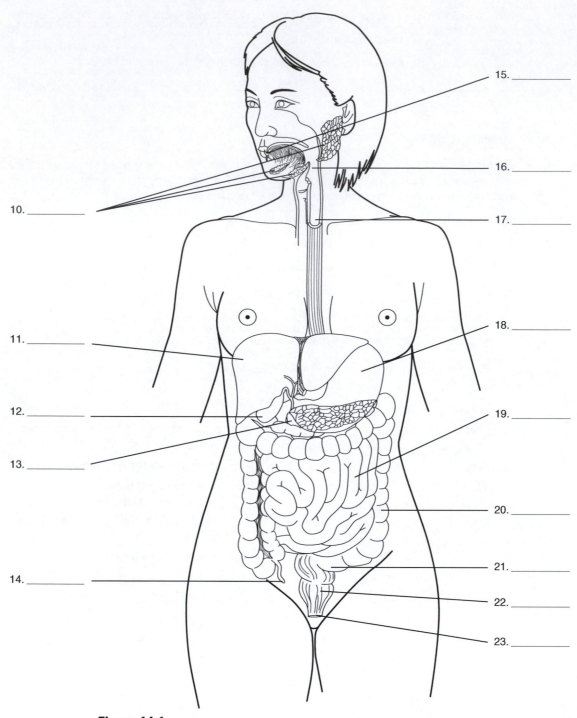

Figure 14.1

Short Answer

24. List the four accessory organs of the digestive system that are not part of the digestive tract.

25. List the five processes that help the digestive system bring nutrients into the body.

26. What are the four main ingredients of saliva and what is the function of each?

27. What are the three functions of the stomach?

28. What are gastric pits? What secretions are found in gastric juice?

29. What are the two main functions of the small intestine?

Labeling

Label each structure indicated in Figure 14.2 and then answer the questions below.

Vein — Lymph vessel

Artery

Nerve

30.

31.

32.

33.

34.

35.

36.

Figure 14.2

37. Which layer is responsible for motility of the digestive tract?

38. Which layer carries blood vessels and nerves?

39. Which layer protects and anchors the digestive tract?

40. Which layer contains smooth muscle?

14.6 Accessory organs aid digestion and absorption

14.7 The large intestine absorbs nutrients and eliminates waste

14.8 How nutrients are absorbed

14.9 Endocrine and nervous systems regulate digestion

Matching

Write the letter(s) of each function next to the correct organs.

_____ 1. **liver**

_____ 2. **pancreas**

_____ 3. **gallbladder**

_____ 4. **colon**

_____ 5. **cecum**

_____ 6. **large intestine**

_____ 7. **stomach**

a. is divided into ascending, transverse, and descending sections

b. stores fat-soluble vitamins

c. secretes sodium bicarbonate to maintain the neutral pH required for its function

d. stores excess glucose as glycogen

e. detoxifies substances

f. a pouch that receives chime from the small intestine

g. absorbs water and produces solid waste

h. produces bile

i. receives nutrients delivered by the hepatic portal system

j. secretes proteases to digest protein and amylase to digest carbohydrates

k. concentrates and stores bile

l. breaks down old red blood cells

m. stretching increases peristalsis

n. delivers digestive enzymes to the duodenum by way of two ducts

o. cholecystokinin (CCK) is produced and stimulates the release of bile

Completion

Complete the table below by filling in the information for each nutrient.

Nutrient	Digestion begins here	Digestion is completed here	Enzymes required for digestion	Most absorption occurs here	Method(s) of transport to body
8. Proteins	a.	b.	c.	d.	e.
9. Carbohydrates	a.	b.	c.	d.	e.
10. Lipids	a.	b.	c.	d.	e.

True/False

Mark each statement as true or false. If a statement is marked false, replace the underlined term or phrase with the term or phrase needed to make the statement true.

_____ 11. The products of lipid digestion are fatty acids and <u>amino acids</u>.

_____ 12. The function of micelles is to transport <u>proteins</u>.

_____ 13. Chylomicrons are clusters of <u>triglycerides</u> coated with <u>proteins</u>.

_____ 14. Stretching of the duodenum <u>decreases</u> segmentation.

_____ 15. Acid in chyme triggers the release of <u>gastrin</u> by the duodenum.

_____ 16. The pancreas increases secretion of digestive enzymes in response to <u>cholecystokinin</u>.

14.10 Nutrition: You are what you eat

14.11 Weight control: Energy consumed versus energy spent

14.12 Disorders of the digestive system

14.13 Eating disorders: Anorexia nervosa and bulimia

Short Answer

1. Why is it healthier to eat complex carbohydrates than simple sugars?

2. What foods provide saturated fats to the diet? Why are saturated fats considered to be unhealthy?

3. Why are eight of the amino acids used in protein synthesis called "essential amino acids"?

4. What is a complete protein? What foods provide complete protein?

5. What percentage of daily calories should come from each of the three main nutrient groups?

6. Why is it unnecessary to take supplements that contain fat-soluble vitamins?

7. What is fiber and why is it not considered a nutrient source?

8. Under the following conditions, will the BMR increase, decrease, or remain the same?

   a. gender differences

   b. stress

   c. eating

   d. not eating

   e. obesity

Completion

Write in the correct digestive system or eating disorder, cause, and symptoms to complete the table below.

Disorder	Cause	Symptoms
Lactose intolerance	9.	10.
11.	small protrusion of the intestinal wall, may be caused by a low-fiber diet	sometimes has no symptoms
12.	liver inflammation caused by contaminated food or water	brief illness
Celiac Disease	13.	14.
Diverticulosis	15.	16.
17.	high levels of cholesterol in bile	18.

Disorder	Cause	Symptoms
19.	excessive dieting, cause unknown	20.
Bulemia	21.	22.

Looking Back/Looking Forward

1. Chapter 3 described various forms of membrane transport. Cholera causes an increased solute concentration in the intestines. Explain why this would lead to dehydration in a cholera patient.

2. Chapter 16 describes the female menstrual cycle and the regulatory role of estrogen, a steroid hormone. Why might an anorexic female experience amenorrhea (absence of the menstrual cycle)?

Chapter Test

Multiple Choice

1. The gastrointestinal tract is composed of
   a. all the organs of the digestive system.
   b. the accessory organs of the digestive system.
   c. the hollow organs of the digestive system.
   d. the blood vessels and nerves that supply the digestive system.

2. Which layer of the gastrointestinal tract wall is connective tissue that contains blood vessels, lymph vessels, and nerves?
   a. mucosa
   b. submucosa
   c. serosa
   d. muscularis

3. Food is propelled forward through the GI tract by
   a. mechanical processing.
   b. peristalsis.
   c. segmentation.
   d. motility.

4. The innermost region of a tooth containing blood vessels and nerves is the
    a. enamel.
    b. crown.
    c. pulp cavity.
    d. dentin.

5. The tongue is primarily
    a. skeletal muscle.
    b. smooth muscle.
    c. connective tissue.
    d. epithelial tissue.

6. Which of the following is *not* a component of saliva?
    a. mucin
    b. bicarbonate
    c. lysozyme
    d. proteinase

7. Which of the following represents the path taken by food after leaving the mouth?
    a. pharynx, larynx, esophagus, small intestine, large intestine
    b. pharynx, esophagus, stomach, small intestine, large intestine
    c. larynx, esophagus, stomach, large intestine, small intestine
    d. larynx, esophagus, stomach, small intestine, large intestine

8. Malfunction of the lower esophageal sphincter may result in
    a. bulimia.
    b. acid reflux.
    c. stomach ulcer.
    d. diarrhea.

9. Gastric glands secrete
    a. HCl and pepsinogen.
    b. gastrin and secretin.
    c. gastrin and lipases.
    d. intrinsic factor and amylase.

10. Enzyme activity in the stomach contributes to the digestion of
    a. carbohydrates.
    b. lipids.
    c. proteins.
    d. all of the above

11. Most of the digestion that takes place in the small intestine occurs in the
    a. duodenum.
    b. ileum.
    c. jejunum.
    d. lacteals.

12. The surface area of the small intestine is greatly increased by the presence of
    a. serosa and submucosa.
    b. muscularis and epithelial cells.
    c. duodenum, jejunum, and ileum.
    d. villi and microvilli.

13. The pancreas secretes enzymes that aid the digestion of
    a. carbohydrates.
    b. lipids.
    c. proteins.
    d. all of the above

14. Bile is produced by the _____ and stored in the _____.
    a. gallbladder, small intestine
    b. gallbladder, liver
    c. liver, small intestine
    d. liver, gallbladder

15. The hepatic portal system delivers nutrient-rich blood from the _____ to the _____.
    a. GI tract, liver
    b. liver, small intestine
    c. stomach, liver
    d. liver, gall bladder

16. The large intestine absorbs primarily
    a. nutrients.
    b. vitamins and minerals.
    c. lipids.
    d. water.

17. Which of the following represents the correct sequence in which structures involved in lipid absorption are formed?
    a. micelles, triglycerides, chylomicrons
    b. triglycerides, chylomicrons, micelles
    c. micelles, chylomicrons, triglycerides
    d. chylomicrons, micelles, triglycerides

18. Proteins and carbohydrates are absorbed from the small intestine by
    a. diffusion.
    b. facilitated diffusion.
    c. active transport.
    d. osmosis.

19. Most digestive regulatory processes involve the
    a. mouth and stomach.
    b. stomach and large intestine.
    c. stomach and small intestine.
    d. small intestine and large intestine.

20. Complex carbohydrates are better nutritionally than simple carbohydrates because
    a. simple carbohydrates contain more fats.
    b. simple carbohydrates contain more calories.
    c. complex carbohydrates provide a more lasting source of energy.
    d. complex carbohydrates are easier to absorb.

21. Essential amino acids are those that
    a. are required for protein synthesis.
    b. the body is capable of synthesizing.
    c. the body is not capable of synthesizing.
    d. are quickly broken down in the body.

22. Water soluble vitamins _____ stored in the liver and _____ get consumed daily.
    a. are, do not
    b. are, do
    c. are not, do not
    d. are not, do

23. Your basal metabolic rate is
    a. the number of calories you consume each day.
    b. the number of calories you use each day.
    c. the number of calories your body needs to perform essential activities.
    d. the number of calories your body needs to perform nonessential activities.

24. To lose 1 pound of fat, you must spend
    a. 1,000 calories.
    b. 1,500 calories.
    c. 2,500 calories.
    d. 3,500 calories.

25. A form of hepatitis transmitted by contaminated food or water is
    a. hepatitis A.
    b. hepatitis B.
    c. hepatitis C.
    d. hepatitis E.

Self-Test

Each of the Key Concepts listed at the beginning of this chapter has been rewritten as a question below. After successfully completing the study guide exercises and the Chapter Test, you should be able to answer each of these questions. Refer to the Key Concepts list at the beginning of this chapter to check your answers.

1. What is the function of the digestive system?

2. What are the four layers in the wall of the gastrointestinal tract?

3. What are sphincters composed of, and what is their function?

4. What are the five basic processes of the digestive system?

5. What are two types of motility that function in the digestive system?

6. What occurs in the mouth?

7. What happens in the pharynx?

8. Which two organs are connected by the esophagus?

9. What are the functions of the stomach?

10. What are the functions of the small intestine?

11. What are the four accessory organs of the digestive system?

12. What are the functions of the pancreas?

13. What are the functions of the liver?

14. What are the functions of the gallbladder?

15. What are the functions of the large intestine?

16. What transport processes are used in the absorption of carbohydrates, proteins, lipids, and water?

17. Which two body systems are involved in the regulation of the digestive system?

18. What are the names of three classes of compounds that are essential nutrients in the diet?

19. How does energy consumption and energy usage affect body weight?

20. What are the eight disorders of the digestive system discussed in this chapter?

15

The Urinary System

Chapter Summary and Key Concepts

After reading and studying this chapter, you should know the following:

Sections 15.1, 15.2, 15.3

1. Excretion refers to processes that remove wastes and excess materials from the body.

2. Excretion is accomplished by the digestive system, the lungs, the skin, the liver, and the urinary system.

3. The urinary system consists of the kidneys, ureters, bladder, and urethra.

4. The urinary system maintains a constant internal environment by regulating water balance and body levels of nitrogenous wastes, ions, and other substances.

5. The main organs of the urinary system are the kidneys.

6. The kidneys form urine and consist of the outer cortex, the inner medulla, and the renal pelvis.

7. The ureters transport urine to the bladder, the bladder stores urine, and the urethra carries urine from the body.

8. The nephron is the functional unit of the kidney.

9. Blood supply to the nephron is provided by the afferent arterioles, peritubular capillaries, and vasa recta.

Sections 15.4, 15.5, 15.6

10. The formation of urine involves three processes: glomerular filtration, tubular reabsorption, and tubular secretion.

11. The glomerular filtrate is produced by high blood pressure in the glomerular capillaries, and it contains water and small solutes in the same concentration as blood plasma.

12. Tubular reabsorption returns filtered water, sodium, major nutrients, and bicarbonate to the blood.

13. Tubular secretion moves substances from the capillaries into the tubules.

14. The kidneys can excrete excess water by producing dilute urine and can conserve water by producing concentrated urine.

Sections 15.7, 15.8

15. The kidneys contribute to homeostasis by adjusting blood volume and pressure, regulating salt balance, maintaining blood pH, controlling production of red blood cells, and participating in the formation of vitamin D.

16. Antidiuretic hormone influences water balance by increasing the water permeability of the collecting duct.

17. Blood volume is influenced by aldosterone, renin, and atrial natriuretic hormone.

18. The kidneys maintain blood pH by secreting H^+.

19. The kidneys stimulate red blood cell production by secreting erythropoietin.

20. Disorders of the urinary system include kidney stones, acute and chronic renal failure, and urinary tract infections.

Exercises

Complete the exercises for each section after you have read and studied the section. If you cannot answer some questions, or if you answer them incorrectly, return to the chapter and review this information. You may find it helpful to work on only one section at a time. When you have completed all sections, take the Chapter Test as an indicator of your mastery of this topic.

15.1 The urinary system contributes to homeostasis

15.2 Organs of the urinary system

15.3 Nephrons produce urine

Define

Write the term(s) for each definition.

_____   1. processes that remove wastes and excess materials from the body

_____   2. ammonia is excreted by the kidneys after it is converted to this molecule

_____   3. a muscular tube that transports urine from the kidney to the bladder

_____   4. the tube that carries urine from the bladder to the body's external opening

_____   5. the functional unit of the kidney that produces urine

_____   6. a network of capillaries surrounded by the glomerular capsule

_____   7. this vessel supplies blood to a nephron

_____ 8. this vessel carries filtered blood away from the glomerulus

_____ 9. the network of capillaries that surround the proximal and distal tubules

_____ 10. these capillaries supply the loop of Henle and the collecting duct

_____ 11. the area of the kidney where urine collects after it is formed

_____ 12. the component of the urinary system that stores urine

Labeling/Coloring

Select different colors for each structure listed below. Color the coding circles and the corresponding structures in Figure 15.1. Structures may appear more than once in the figure.

○ kidney   ○ bladder   ○ nephron

○ renal pelvis   ○ ureter   ○ inferior vena cava

○ renal vein   ○ medulla   ○ collecting duct

○ renal artery   ○ cortex

○ aorta   ○ urethra

Figure 15.1

Short Answer

13. Which components of the urinary system transport and store urine until it is eliminated from the body?

14. How is urine moved through the ureters?

15. List the four regions of a nephron tubule in the order that filtrate passes through during the formation of urine.

16. What is the function of the peritubular capillaries?

17. What is the first step in urine formation?

18. How is the urethra different in males and females?

Labeling

Use the terms below to label Figure 15.2.

glomerulus   loop of Henle   peritubular capillaries

efferent arteriole   distal tubule   afferent arteriole

glomerular capsule   descending limb   collecting duct

ascending limb   proximal tubule

Figure 15.2

15.4   Formation of urine—filtration, reabsorption, and secretion

15.5   The kidneys can produce dilute or concentrated urine

15.6   Urination depends on a reflex

Fill-in-the-Blank

Referenced sections are in parentheses.

1. Modified epithelial cells, called _____, are located on the outer surface of the glomerular capillaries. (15.4)

2. The driving force for glomerular filtration is high _____ _____ in the glomerular capillaries. (15.4)

3. Compared to most other capillaries, the capillaries of the glomerulus are more permeable to _____ and small _____, and less permeable to large _____. (15.4)

4. Water reabsorption in the proximal tubules occurs through water channels in membrane proteins called _____. (15.4)

5. The proximal tubule secretes _____ and _____. (15.4)

6. The distal tubule secretes _____.

7. The ability to form dilute urine depends partly on the _____ of solutes in the _____ _____. (15.5)

8. The solute concentration of the renal medulla is kept high by the recycling of _____. (15.5)

9. The movement of fluid in opposite directions in the ascending and descending limbs of the loop of Henle is called _____ flow. (15.5)

10. Urination depends on a neural reflex called the _____ _____. (15.6)

11. The internal urethral sphincter consists of _____ muscle, and the external urethral sphincter consists of _____ muscles. (15.6)

Short Answer

12. What three processes are involved in the formation of urine?

13. Where are podocytes located, and what is their function?

14. What substances are almost completely reabsorbed during tubular reabsorption?

15. What happens during tubular secretion, and what membrane transport processes are involved?

16. What would cause the kidneys to produce more concentrated urine?

17. What is the effect of ADH on the kidney?

Critical Thinking

18. Why does protein in the urine signal possible kidney damage?

Completion

For each characteristic listed below indicate the region of the nephron where it occurs. Nephron regions are labeled a–f, and may be used more than once. Some characteristics also occur in more than one region.

Region

a. proximal tubule

b. descending limb of the loop of Henle

c. first portion of the ascending limb of the loop of Henle

d. final portion of the ascending limb of the loop of Henle

e. distal tubule

f. collecting duct

Characteristic

_____ 19. impermeable to water

_____ 20. water permeability in this region is influenced by ADH

_____ 21. NaCl is actively transported

_____ 22. impermeable to NaCl

_____ 23. sodium is actively transported and chlorine moves by diffusion

_____ 24. water moves by simple diffusion

15.7 The kidneys maintain homeostasis in many ways

15.8 Disorders of the urinary system

Word Choice

Circle the term that correctly completes each sentence.

1. ADH is secreted into the (blood/kidney tubules) by the posterior pituitary when blood solute concentration (decreases/increases).

2. In response to ADH, the permeability of the collecting duct to water (decreases/increases), resulting in a more (dilute/concentrated) urine.

3. When blood solute concentration decreases, the (hypothalamus/kidney) signals the posterior pituitary to (decrease/increase) ADH secretion.

4. Aldosterone is produced by the (hypothalamus/posterior pituitary/adrenal gland).

5. Aldosterone (decreases/increases) the reabsorption of (sodium/chloride/water) across the distal tubule and collecting duct.

6. A decrease in blood volume triggers a(n) (decrease/increase) in aldosterone secretion.

7. A decrease in blood pressure causes (decreased/increased) secretion of rennin by cells in the (afferent/efferent) arteriole.

8. The effect of rennin on (angiotensinogen/angiotensin II) is the production of angiotensin I.

9. Angiotensin II stimulates the secretion of (ADH/aldosterone).

10. ANH is a hormone secreted by the (hypothalamus/heart/kidney) that causes (decreased/increased) Na^+ excretion in the urine.

True/False

Mark each statement as true or false. If you mark a statement as false, replace the underlined term or phrase with the term or phrase needed to make the statement true.

_____ 11. The kidneys are considered to be endocrine organs because they secrete <u>hormones</u>.

_____ 12. The kidneys, the <u>hypothalamus</u>, and the posterior pituitary work together to maintain water balance in the body.

_____ 13. An increase in blood volume leads to a <u>decrease</u> in blood pressure.

_____ 14. <u>Renin</u> is secreted by afferent arteriole cells of the juxtaglomerular apparatus.

_____ 15. The kidneys help to maintain acid/base balance by secreting <u>H+</u> into the tubular lumen.

_____ 16. The kidneys regulate the production of <u>red blood cells</u> through the <u>secretion of the hormone erythropoietin</u>.

_____ 17. Erythropoietin is released by the kidneys when oxygen levels <u>increase</u>.

_____ 18. The kidneys are important in activating Vitamin D, which is required for the absorption of <u>sodium and chloride</u> from the digestive tract.

_____ 19. Vitamin D synthesis begins in the <u>skin</u> with exposure to sunlight.

_____ 20. Before reaching the kidneys, the inactive form of Vitamin D is transported to the <u>liver</u>.

_____ 21. Inactive Vitamin D is converted to an active form in the kidneys under the influence of <u>ADH</u>.

Short Answer

22. What causes kidney stones and how are they treated?

23. What is the most common cause of urinary tract infections? Why are UTIs more common in women than in men?

24. How is acute renal failure different from chronic renal failure?

25. What are the benefits and risks of CAPD?

Looking Back/Looking Forward

1. Refer to Fig. 3.30 in Chapter 3 to answer the following questions:
   a) The metabolism of protein produces amino acids, and the breakdown of these amino acids includes the production of ammonia (NH3). Is the metabolism of protein anabolic or catabolic?

   b) Ammonia is eliminated as a waste product by the kidneys; how are the other products of protein metabolism used by the body?

2. Kidney cancer is one of the ten most common cancers, as discussed in Chapter 18, and can be caused by ingestion of toxic chemicals. Consider the role of the kidneys in concentrating urine to explain why the kidneys are one of the body organs most likely to be affected by toxins.

Chapter Test

Multiple Choice

1. Which of the following is(are) not part of the urinary system?
   a. kidneys
   b. liver
   c. bladder
   d. ureters

2. Which of the following statements about urea is false?
   a. It is produced in the liver.
   b. It is synthesized from ammonia.

    c. It is a by-product of protein metabolism.

    d. It is metabolized and excreted through the lymph.

3. The functional unit of the kidney is the

    a. renal medulla.

    b. pyramid.

    c. nephron.

    d. glomerulus.

4. The structure of the glomerular capillaries contributes to the formation of filtrate because

    a. the walls of the capillaries are strengthened by smooth muscle to withstand the high blood pressure in the glomerulus.

    b. the walls of the capillaries contain tight junctions, preventing excess plasma from leaking into the glomerulus.

    c. the walls of the capillaries are much more permeable to water and small solutes than other capillaries.

    d. the branching of the capillaries provides for even distribution of blood pressure.

5. Which of the following *incorrectly* describes the process of urine formation?

    a. Glomerular filtration—fluid filtered from the blood in the glomerulus moves into the glomerular capsule.

    b. Tubular reabsorption—fluid filtered from the proximal tubule is absorbed by the distal tubule.

    c. Tubular secretion—essential solutes are secreted from the tubules into the peritubularcapillaries.

    d. b and c are incorrect.

    e. a, b, and c are incorrect.

    f. a, b, and c are all correctly described.

6. Which of the following *correctly* summarizes how the process of urine formation contributes to the final urine content?

    a. glomerular filtration – tubular reabsorption + tubular secretion = excretion

    b. glomerular filtration + tubular secretion + tubular reabsorption = excretion

    c. tubular reabsorption + glomerular filtration – tubular secretion = excretion

    d. tubular secretion – glomerular filtration + tubular reabsorption = excretion

7. Blood pressure is the driving force that moves fluid out of all the capillaries in the body. Why is so much more fluid moved out of the capillaries in the glomerulus of the kidneys than in other areas of the body?

    a. Movement of fluid out of the glomerular capillaries is not opposed by osmotic pressure.

    b. Blood entering the kidneys contains more water than capillary blood in other parts of the body.

    c. The blood pressure in the glomerular capillaries is twice as high as in other capillaries of the body.

    d. Fluid is actively transported out of the glomerular capillaries.

8. The _____ arteriole carries blood into the glomerulus, and the _____ arteriole carries blood away.
   a. proximal, distal
   b. ascending, descending
   c. filtering, collecting
   d. afferent, efferent

9. Substances that are reabsorbed from the proximal tubule will enter the
   a. afferent arteriole.
   b. peritubular capillaries.
   c. collecting duct.
   d. descending arteriole.

10. Most tubular reabsorption occurs in the _____, where lumen cells contain _____.
    a. proximal tubule, microvilli
    b. loop of Henle, active transport proteins
    c. collecting duct, pores
    d. renal medulla, proton pumps

11. Which of the following is not almost completely reabsorbed from the filtrate?
    a. glucose
    b. water
    c. creatinine
    d. salt

12. Sodium is reabsorbed from the proximal tubule by
    a. passive diffusion.
    b. facilitated diffusion.
    c. active transport.
    d. osmosis.

13. Which two ions are secreted by the proximal tubule in maintaining acid-base balance?
    a. hydrogen and bicarbonate
    b. urea and ammonium
    c. hydrogen and ammonium
    d. potassium and hydrogen

14. The diffusion of water out of the descending limb creates a gradient that favors
    a. the diffusion of water into the ascending limb.
    b. the diffusion of NaCl out of the ascending limb.
    c. the active transport of glucose out of the ascending limb.
    d. the active transport of H^+ into the descending limb.

15. The action of ADH on the collecting duct produces
    a. a dilute urine.
    b. a concentrated urine.
    c. urine with a high glucose content.
    d. the presence of protein in the urine.

16. Which urethral sphincter is voluntary?
    a. internal
    b. external
    c. secondary
    d. primary

17. The juxtaglomerular apparatus consists of cells of the
    a. afferent arteriole and distal tubule.
    b. afferent arteriole and proximal tubule.
    c. efferent arteriole and distal tubule.
    d. efferent arteriole and collecting duct.

18. Caffeine acts as a mild diuretic because it
    a. stimulates the release of ADH.
    b. stimulates sodium reabsorption.
    c. inhibits sodium reabsorption.
    d. increases the permeability of the glomerular capillaries.

19. The effect of aldosterone on the distal tubule and collecting duct is to
    a. increase reabsorption of sodium.
    b. decrease reabsorption of sodium.
    c. increase excretion of water.
    d. increase secretion of H^+.

20. The greatest stimulus for release of aldosterone is
    a. high sodium concentrations in the filtrate.
    b. decreased blood volume.
    c. increased blood volume.
    d. increased release of renin.

21. Renin is produced by
    a. afferent arteriole cells in the juxtaglomerular apparatus.
    b. the liver.
    c. the pituitary.
    d. efferent arteriole cells in the juxtaglomerular apparatus.

22. The kidneys regulate the production of red blood cells by secreting
    a. renin.
    b. erythropoietin.
    c. angiotensin I.
    d. vitamin D.

23. If the liver were unable to produce angiotensinogen, which of the following might occur?
    a. the kidneys would be unable to produce renin
    b. ADH levels would increase
    c. blood vessels would dilate
    d. angiotensin II would not be available

24. What is the role of the kidneys in the formation of vitamin D?
    a. The kidneys function alone in the synthesis of vitamin D.
    b. The kidneys function with the skin and the liver in vitamin D synthesis.
    c. The kidneys increase vitamin D secretion when blood pressure is low.
    d. The kidneys secrete vitamin D and erythropoietin together.

25. Which of the following are characteristics of acute renal failure?
    a. short-term duration
    b. incurable
    c. end-stage renal failure
    d. associated with Type I diabetes

Self Test

Each of the Key Concepts listed at the beginning of this chapter has been rewritten as a question below. After successfully completing the study guide exercises and the Chapter Test, you should be able to answer each of these questions. Refer to the Key Concepts list at the beginning of this chapter to check your answers.

1. What is excretion?

2. What systems and organs contribute to excretion?

3. What are the organs of the urinary system? What are the functions of the urinary system?

4. How does the urinary system contribute to homeostasis?

5. What are the main organs of the urinary system?

6. What are the three main regions of the kidney?

7. How do the ureters, the bladder, and the urethra contribute to the function of the urinary system?

8. What is the functional unit of the kidney?

9. What vessels supply blood to a nephron?

10. What are the three processes involved in the formation of urine?

11. What pressure contributes to the formation of glomerular filtrate, and what does the filtrate contain?

12. What occurs in tubular reabsorption?

13. What occurs in tubular secretion?

14. What changes in water excretion do the kidneys make to produce dilute or concentrated urine?

15. What are five ways in which the kidneys contribute to homeostasis?

16. How does ADH influence water balance?

17. What are three hormones or chemicals that influence blood volume?

18. How do the kidneys maintain blood pH?

19. How do the kidneys stimulate red blood cell production?

20. What four disorders of the urinary system are discussed in this chapter?

16

Reproductive Systems

Chapter Summary and Key Concepts

After reading and studying this chapter, you should know the following:

Section 16.1

1. The reproductive system consists of primary reproductive organs, accessory glands, and ducts.

2. The primary organs of the male reproductive system are the testes, which secrete testosterone and produce sperm.

3. Sperm are produced in the seminiferous tubules and mature in the epididymis.

4. Sperm leave the body through the vas deferens and urethra in a fluid called semen.

5. Semen contains sperm and glandular secretions that provide nutrients, alkaline fluid, and mucus.

6. Sperm cells contain 23 chromosomes and are produced by a cell-division process called meiosis.

7. Testosterone, LH, and FSH regulate the male reproductive system.

Section 16.2

8. The primary organs of the female reproductive system are the ovaries, which release oocytes.

9. The oviduct is a hollow tube that transports the oocyte from the ovary to the uterus, and it is the site of fertilization.

10. The uterus supports the development of a fertilized egg, and it consists of an inner endometrium and an outer myometrium.

11. The vagina is the female organ of sexual intercourse and is the birth canal.

12. The mammary glands are accessory organs that produce and store milk.

Sections 16.3, 16.4, 16.5

13. The ovaries and uterus undergo a monthly pattern of changes called the menstrual cycle, which consists of an ovarian cycle and a uterine cycle.

14. The ovarian cycle results in the release of a secondary oocyte capable of being fertilized.

15. The uterine cycle prepares the endometrium of the uterus to accept and nurture a fertilized egg.

16. GnRH, LH, FSH, estrogen, and progesterone are hormones involved in regulating the female reproductive system.

17. The human sexual response is a series of events that coordinate sexual function to accomplish intercourse and fertilization.

18. Birth control methods are designed to prevent fertilization.

19. Disorders of the reproductive system include infertility, complications of pregnancy, cancers and tumors of the reproductive organs, and sexually transmitted diseases.

20. STDs caused by bacteria are curable, while those caused by viruses are not. STDs include gonorrhea, syphilis, chlamydia, AIDS, hepatitis B, herpes, and genital warts.

Exercises

Complete the exercises for each section after you have read and studied the section. If you cannot answer some questions, or if you answer them incorrectly, return to the chapter and review this information. You may find it helpful to work on only one section at a time. When you have completed all sections, take the Chapter Test as an indicator of your mastery of this topic.

16.1 Male reproductive system delivers sperm

Labeling

Label each of the structures indicated in Figure 16.1.

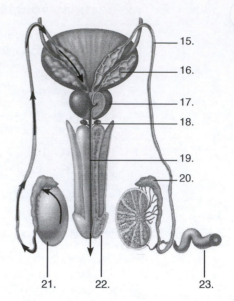

Figure 16.1

1. _____
2. _____
3. _____
4. _____
5. _____
6. _____
7. _____
8. _____
9. _____
10. _____
11. _____
12. _____

13. _____
14. _____
15. _____
16. _____
17. _____
18. _____
19. _____
20. _____
21. _____
22. _____
23. _____

Fill-in-the-Blank

24. Sperm are the male _____.

25. In humans, diploid cells have _____ chromosomes and haploid cells have _____ chromosomes.

26. Secretion of FSH by the anterior pituitary is inhibited by the hormone _____.

27. Blood levels of testosterone are maintained by a _____ feedback loop involving the hypothalamus and the _____ _____.

28. Decreased levels of testosterone stimulate the hypothalamus to secrete

_____.

29. _____ released by the pituitary increases secretion of testosterone.

30. When the temperature of developing sperm is too low, the _____ brings the testes nearer to the body.

31. Secretions from the _____ _____ may help to induce muscle contraction in the female reproductive tract.

32. Sperm leave the male body in a fluid called _____.

33. Once sperm production begins at _____, sperm will be produced_____.

34. The seminiferous tubules contain _____ cells that nourish developing sperm.

Completion

Write the structure responsible for each function listed below.

_____ 35. production of sperm and testosterone

_____ 36. keeps the testes at the proper temperature

_____ 37. location within the testes where sperm are produced

_____ 38. site of sperm maturation and storage

_____ 39. transports sperm out of the epididymis

_____ 40. transports sperm from the vas deferens to the urethra

_____ 41. transports sperm outside the male body

_____ 42. secretes fructose and most of the seminal fluid

_____ 43. gland that adds an alkaline fluid to the semen

_____ 44. secretes lubricating mucus

Labeling

In Figure 16.2a, label each type of cell indicated. In Figure 16.2b, label each region of a mature sperm. Then, answer the questions that follow.

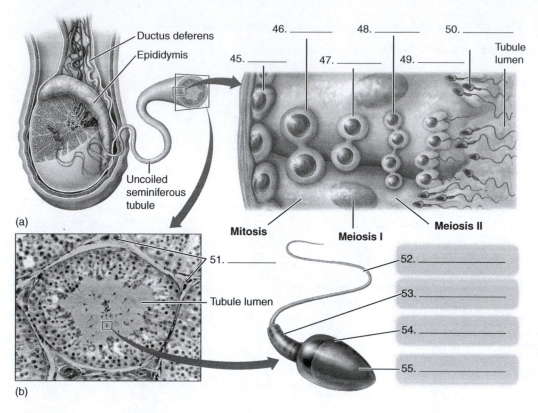

Figure 16.2

56. Are spermatogonia haploid or diploid?

57. Are sperm haploid or diploid?

58. What type of cell division produces primary spermatocytes?

59. What type of sperm cell is produced by meiosis I?

60. What type of sperm cell is produced by meiosis II?

61. Are spermatids capable of fertilizing an egg?

62. What is contained in the head of a mature sperm?

63. What is contained in the acrosome?

64. What is the function of the midpiece?

65. What is the function of the tail?

16.2 Female reproductive system produces eggs and supports pregnancy

Define

Write the term(s) of the female reproductive system for each definition.

_____ 1. the female reproductive organs

_____ 2. organ that secretes estrogen and progesterone

_____ 3. tube leading from the ovary to the uterus

_____ 4. fingerlike projections on the fallopian tubes that help to move the oocyte

_____ 5. organ where a fertilized egg grows and develops

_____ 6. the lower part of the uterus through which sperm enter

_____ 7. layer of the uterus that helps to form the placenta during pregnancy

_____ 8. the birth canal

_____ 9. external female genitalia

_____ 10. during early development this organ originates from the same tissue as the penis

_____ 11. modified sweat glands responsible for lactation

Short Answer

12. How many eggs are released by the female reproductive system each month?

13. Where does fertilization occur?

14. Describe the two layers of the uterus.

Labeling/Coloring

Select different colors for each structure listed below. Color the coding circles and the corresponding structures in Figure 16.3.

○ clitoris   ○ uterus

○ vagina   ○ labium minora

○ labium majora   ○ oviduct

○ ovary   ○ fimbriae

○ bladder   ○ urethra

○ cervix

Figure 16.3

16.3 The menstrual cycle consists of ovarian and uterine cycles

16.4 Human sexual response, intercourse, and fertilization

16.5 Birth control methods: Controlling infertility

Completion

Use the terms below to complete the paragraph, and then review the completed paragraph for accuracy.

endometrium	secondary	pellucida
ovulation	ovarian	meiosis II
corpus	Graafian	antrum
anterior	LH	puberty
FSH	ovarian	follicle
decrease	granulosa	primary
meiosis I	zona	pituitary
ovulation	GnRH	hypothalamus
progesterone	luteum	glycoprotein

Meiosis I

The (1)_____ cycle involves the maturation and release of oocytes

from the ovary. Each ovary contains approximately one million

(2)_____ oocytes at birth. The oocytes have already begun the process

of (3)_____. Continued development of an oocyte does not begin until

(4)_____, and an oocyte will not complete (5)_____ until

a sperm enters.

Each primary oocyte is surrounded by nourishing cells called

(6)_____ cells. These cells, along with the oocyte, make up an imma-

ture (7)_____. About a dozen immature follicles begin the

development process each month in response to (8)_____ and LH

released by the (9)_____ (10)_____. At the beginning

of the ovarian cycle, (11)_____ from the (12)_____

increases the concentration of these hormones, triggering oocyte development.

Usually only one oocyte will complete the maturation process and be released dur-

ing (13)_____.

As an immature follicle grows, the granulosa cells begin to secrete

(14)_____ that become a noncellular coating called the

(15)_____ (16)_____. Next, a fluid-filled space called the

(17)_____ develops, and the follicle begins secreting estrogen and

(18)_____. The primary ooctye will now finish the process of

(19)_____, producing a (20)_____ oocyte and a polar

body. This mature follicle is now called a (21)_____ follicle.

As estrogen (produced by the follicle) levels rise, the pituitary is stimulated to

release a surge of the hormone (22)_____. This hormone surge

triggers the process of (23)_____, the follicle ruptures, and the
secondary oocyte is released. The remnants of the follicle will now form the
(24)_____ (25)_____, which secretes large amounts of
estrogen and progesterone to prepare the (26)_____ for implantation,
if fertilization should occur. If fertilization does not occur, the corpus luteum degen-
erates, estrogen and progesterone levels (27)_____, and the
(28)_____ cycle will begin again.

Short Answer

Refer to Figure 16.4, describing the menstrual cycle, to answer the following questions.

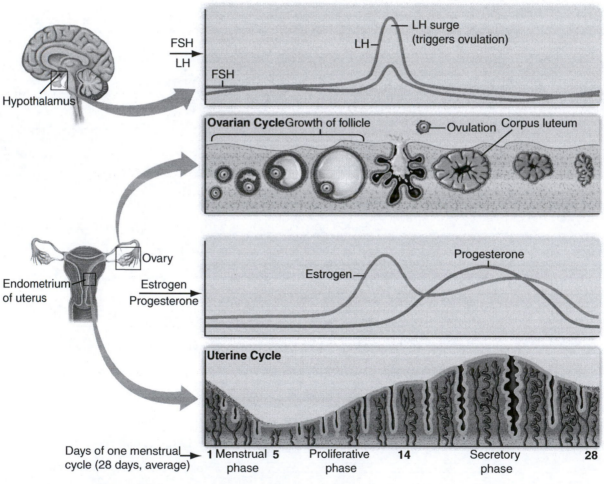

Figure 16.4

29. Which days constitute the menstrual phase?

30. What body levels of estrogen and progesterone contribute to the events of the menstrual phase?

31. Which days constitute the proliferative phase?

32. What hormone activity triggers the proliferative phase?

33. What event of the ovarian cycle coincides with the end of the proliferative phase?

34. What uterine phase covers the last half of the menstrual cycle?

35. What hormone level is the highest during the last half of the menstrual cycle?

36. What event is the uterus preparing for during this phase?

37. In the absence of fertilization, what event in the ovarian cycle causes the levels of estrogen and progesterone to decline?

38. Does positive feedback influence the menstrual cycle before or after ovulation?

39. Describe the positive feedback event that influences the menstrual cycle.

40. What prevents development of a second follicle during the luteal phase?

Completion

Name each birth control method described in the table.

Method of Birth Control	Description
41.	the vas deferens is tied in two places and the segment between the ties is removed
42.	consists of an injection of progesterone that lasts 3 months, inhibiting FSH and LH

(Continued)

Method of Birth Control	Description
43.	a chemical in the form of a foam, cream, jelly, or douche that destroys sperm
44.	a plastic or metal piece inserted into the uterus that causes inflammation, interfering with implantation
45.	intercourse is avoided from about 5 days before ovulation until 3 days after ovulation
46.	a monthly injection of estrogen and progesterone
47.	a patch that slowly releases hormones into the bloodstream
48.	blocks the action of progesterone, causing a chemical abortion

Critical Thinking

49. The female egg is large compared to the male sperm, and the number of eggs produced by the female is small compared to the number of sperm produced by the male. Explain why, structurally and functionally, this occurs.

16.6 Infertility: Inability to conceive

16.7 Sexually transmitted diseases: A worldwide concern

Short Answer

1. What is the definition of infertility?

2. What is the minimum sperm count required for male fertility?

3. What is the most common cause of female infertility?

4. Briefly describe the process of *in vitro* fertilization (IVF).

5. How does gamete intrafallopian transfer (GIFT) differ from IVF?

Completion

For each complication or disease of the reproductive system listed below, indicate the cause, symptoms, and treatment.

6. Gonorrhea
   a. cause:

   b. symptoms:

   c. treatment:

7. Syphilis
   a. cause:

   b. symptoms:

   c. treatment:

8. Chlamydia
   a. cause:

   b. symptoms:

   c. treatment:

9. HIV
   a. cause:

   b. symptoms:

   c. treatment:

10. Hepatitis B
    a. cause:

    b. symptoms:

    c. treatment:

11. Genital herpes

   a.   cause:

   b.   symptoms:

   c.   treatment:

12. HPV

   a.   cause:

   b.   symptoms:

   c.   treatment:

Looking Back/Looking Forward

1.   Chapter 9 discussed the immune system. Human sperm have a cell surface marker that is universally recognized by the immune system and prevents attack by immune cells. Why is this important to the survival of sperm as they enter the female reproductive tract?

2.   Chapter 19 discusses genetics and inheritance; individuals with an X and a Y chromosome are genetically male. The male and female reproductive tracts are identical early in development, with the presence of testosterone triggering development as a male. Sometimes errors occur, and the tissues are either insensitive to testosterone or too little testosterone is produced. How might this affect development of the male reproductive tract?

Chapter Test

Multiple Choice

1.   Which of the following is a secondary sexual characteristic?
   a.   testosterone
   b.   testis
   c.   seminal vesicle
   d.   muscle mass

2. The male gamete is the _____, and the female gamete is the _____.
   a. testis, ovary
   b. testosterone, estrogen
   c. vas deferens, oviduct
   d. sperm, egg

3. Which of the following is *not* a characteristic of Sertoli cells?
   a. located between the seminiferous tubules
   b. secrete inhibin
   c. stimulated by testosterone
   d. support developing sperm

4. The epididymis
   a. is a tube connecting the vas deferens to the urethra.
   b. delivers secretions of the seminal vesicles into the urethra.
   c. is a coiled duct just outside the testis.
   d. is a mass of tightly coiled tubules where sperm are produced.

5. From production to exiting the male body, the pathway of a sperm is
   a. epididymis, seminiferous tubules, vas deferens, ejaculatory duct, urethra.
   b. vas deferens, ejaculatory duct, seminiferous tubules, epididymis, vas deferens.
   c. seminiferous tubules, vas deferens, epididymis, urethra, ejaculatory duct.
   d. seminiferous tubules, epididymis, vas deferens, ejaculatory duct, urethra.

6. In the process of sperm development, secondary spermatocytes develop into _____, which mature to become _____.
   a. primary spermatocytes, spermatids
   b. primary spermatocytes, sperm
   c. spermatids, sperm
   d. spermatids, spermatogonia

7. Which region of the sperm is *incorrectly* described?
   a. midpiece—carries secretions from the male glands that help to neutralize acidity in the vagina
   b. head—contains the male chromosomes
   c. acrosome—contains enzymes to aid penetration of the egg
   d. tail—provides for movement of the sperm through the female reproductive tract

8. Spermatogonia are _____ and sperm are _____.
   a. haploid, haploid
   b. haploid, diploid
   c. diploid, haploid
   d. diploid, diploid

9. Sperm production
   a. is controlled by a positive feedback loop.
   b. occurs in the seminiferous tubules.
   c. is stimulated by inhibin.
   d. begins shortly after birth and continues through old age.

10. The secretion of testosterone is initiated by _____ released by the hypothalamus.
    a. GnRH
    b. LH
    c. FSH
    d. ADH

11. If the pathway from the uterus to the oviduct is blocked,
    a. an oocyte cannot be released.
    b. an oocyte cannot enter the oviduct.
    c. fertilization cannot occur.
    d. the uterine cycle will be disrupted.

12. The _____ provide(s) the muscular force to expel a mature fetus during labor and birth.
    a. fimbria
    b. endometrium
    c. myometrium
    d. cervix

13. At birth, the ovary
    a. already contains approximately one million secondary oocytes.
    b. produces high levels of LH to prevent maturation of follicles until puberty.
    c. is controlled by inhibin released by the anterior pituitary.
    d. contains many more immature follicles than will be present at puberty.

14. Which of the following is *not* released from the ovary with the secondary oocyte?
    a. the polar body
    b. the zona pellucida
    c. the granulosa cells
    d. the corpus luteum

15. The function of the corpus luteum is to
    a. protect the secondary oocyte from fertilization by multiple sperm.
    b. nourish the secondary ooctye until it reaches the uterus.
    c. secrete progesterone and estrogen to prepare the uterus for implantation.
    d. secrete progesterone and estrogen to prevent release of a second egg before the end of the uterine cycle.

16. Menstruation is triggered by
    a. the implantation of a fertilized egg.
    b. the process of ovulation.
    c. a decline in the levels of progesterone and estrogen.
    d. an increase in the levels of progesterone and estrogen.

17. If fertilization does not occur, changing hormone levels leading to menstruation will result from
    a. the absence of a chorion.
    b. the degeneration of the corpus luteum.
    c. the growth of a new follicle.
    d. proliferation of the endometrium.

18. The first half of the menstrual cycle is influenced by positive feedback. This occurs when
    a. rising levels of estrogen in the first half of the cycle trigger a surge in LH.
    b. high levels of estrogen and progesterone inhibit LH and FSH.
    c. decreasing levels of estrogen stimulate an increase in LH.
    d. decreasing levels of progesterone stimulate an increase in LH and FSH.

19. Erection occurs when
    a. arterioles leading into the vascular compartments of the penis dilate, while veins draining the penis are compressed.
    b. arterioles leading into the vascular compartment of the penis constrict, while veins draining the penis dilate.
    c. muscular contractions of the ejaculatory duct compress the vessels that supply the vascular compartments of the penis.
    d. smooth muscle in the penis contracts due to stimulation by the sympathetic nervous system.

20. Hormonal methods of birth control administer progesterone and estrogen in amounts that
    a. stimulate the release of GnRH.
    b. stimulate the release of LH and FSH.
    c. inhibit the release of LH and FSH.
    d. prevent the proliferative phase of the uterine cycle.

21. One of the drawbacks in using a diaphragm is
    a. the insertion of the diaphragm must be done in a doctor's office.
    b. a high failure rate.
    c. that diaphragms do not provide protection against STDs.
    d. the side effects, which may include headaches and nausea.

22. Male infertility usually results from
    a. unusually high levels of testosterone.
    b. failure of the testis to descend completely into the scrotum.
    c. insufficient numbers of normal sperm.
    d. adrenal tumors resulting in increased estrogen secretion.

23. Endometriosis is a condition in which
    a. the endometrium fails to proliferate during the uterine cycle.
    b. the endometrium is insensitive to estrogen and progesterone.
    c. there is abnormal thickening of the myometrium, creating pressure on the endometrium.
    d. endometrial tissue migrates up the oviduct and grows on other organs.

24. Which of the following types of STDs is *incorrectly* matched with its causative organism?
    a. gonorrhea—bacteria
    b. Hepatitis B—virus
    c. HPV—virus
    d. syphilis—virus

25. Hepatitis B
    a. is caused by a bacterium.
    b. can be cured with antibiotics.
    c. can be easily passed through non-sexual contact.
    d. is more contagious than HIV.
    e. a and b

Self-Test

Each of the Key Concepts listed at the beginning of this chapter has been rewritten as a question below. After successfully completing the study guide exercises and the Chapter Test, you should be able to answer each of these questions. Refer to the Key Concepts list at the beginning of this chapter to check your answers.

1. What are the general structures of a reproductive system?

2. What are the primary organs of the male reproductive system, and what are their functions?

3. Where are sperm produced, and where do they mature?

4. What tubular structures carry the sperm outside the body? What fluid contains the sperm?

5. What is contained in semen?

6. How many chromosomes are there in a sperm cell? What type of cell division produces sperm?

7. What are three hormones important in regulating the male reproductive system?

8. What are the primary organs of the female reproductive system, and what is their function?

9. What is the oviduct, and what is its function? Where does fertilization occur?

10. What is the role of the uterus? Name the two layers of the uterus.

11. What are the two functions of the vagina?

12. What is the function of the mammary glands?

13. What are the two organs involved in the menstrual cycle?

14. Which cycle results in the release of a secondary oocyte?

15. Which cycle prepares the endometrium to accept and nurture a fertilized egg?

16. What are the five hormones involved in regulating the female reproductive cycle?

17. The human sexual response coordinates sexual function to accomplish what two events?

18. Birth control methods are designed to prevent what general event?

19. What are the four general classes of reproductive disorders?

20. Which STDs are curable and which are not curable? List several STDs of each type.

17

Cell Reproduction and Differentiation

Chapter Summary and Key Concepts

After reading and studying this chapter, you should know the following:

Sections 17.1, 17.2

1. The cell cycle consists of two phases: interphase and the mitotic phase.

2. Interphase is a growth phase consisting of three stages: G1, S, and G2. During G1, cells grow; during S, the deoxyribonucleic acid (DNA) is replicated; and during G2, cells continue to grow in preparation for cell division.

3. The mitotic phase consists of a nuclear division called mitosis, and a cytoplasmic division called cytokinesis.

4. Human cells contain DNA arranged into 46 chromosomes; each chromosome includes segments of DNA, called genes, that encode a particular protein.

5. Replication is the process of copying a cell's DNA, produces replicated chromosomes that consist of two sister chromatids joined by a centromere, and occurs during the S phase of the cell cycle.

6. Transcription and translation occur when the information in a DNA gene is used to make a protein.

7. Transcription occurs in the nucleus and makes an RNA copy of a DNA gene.

8. Translation occurs in the cytoplasm and results in production of a protein.

9. Three types of RNA participate in translation: messenger RNA carries instructions about how to build the protein, transfer RNA transports amino acids to the growing protein, and ribosomal RNA forms the ribosome.

10. A mutation is a permanent change in the DNA. Mutations may be helpful or harmful, and if not repaired, they may contribute to the process of evolution.

Sections 17.3, 17.4, 17.5

11. Mitosis can be divided into a series of phases called prophase, metaphase, anaphase, and telophase.

12. Mitosis produces two diploid daughter cells that are identical to the parent cell; meiosis produces four haploid cells that are genetically different from the parent cell.

13. Meiosis consists of two successive nuclear divisions called meiosis I and meiosis II.

14. In a female, meiosis results in the production of one functional egg. In a male, meiosis results in the production of four functional sperm.

15. Control of the cell cycle occurs through regulatory proteins called cyclins and the use of cell cycle checkpoints.

16. Environmental factors and selective gene expression lead to the development of specialized cells.

17. Differentiation is the process whereby cells become different from each other.

Sections 17.6, 17.7

18. Reproductive cloning makes a copy of an entire organism and is accomplished by embryo splitting or somatic cell nuclear transfer.

19. Embryo splitting produces multiple organisms that are identical to each other; somatic cell nuclear transfer produces a clone of an adult organism.

20. Therapeutic cloning is done to produce specific cells or tissues for medical treatment.

Exercises

Complete the exercises for each section after you have read and studied the section. If you cannot answer some questions, or if you answer them incorrectly, return to the chapter and review this information. You may find it helpful to work on only one section at a time. When you have completed all sections, take the Chapter Test as an indicator of your mastery of this topic.

17.1 The cell cycle creates new cells

17.2 Replication, transcription, and translation—an overview

Completion

Use the letters in the list of characteristics and in Figure 17.1 to complete each phase of the cell cycle in the table that follows. Some characteristics may be used more than once.

Characteristics

a. includes mitosis and cytokinesis

b. the cell is at its smallest size

c. lasts 7–8 hours in mammalian cells

d. a phase in which healthy cells are not dividing

e. DNA is replicated

f. cell growth is occurring

g. DNA appears as chromatin

h. lasts 30–45 minutes in mammalian cells

i. sister chromatids are produced

j. includes all stages of the cell cycle that occur between cell divisions

k. the cell prepares for cell division

Figure 17.1

Phase	Diagram Labeled Phase	Characteristics
G_0	1. not labeled	2.
G_1	3.	4.
S	5.	6.
G_2	7.	8.
M	9.	10.
Interphase	11.	12.

Fill-in-the-Blank

Referenced sections are in parentheses.

13. The division of the nucleus occurs during the _____

    _____. (17.1)

14. The division of the cytoplasm is called _____. (17.1)

15. Inside the nucleus of human cells, DNA is organized into 46 separate structures called

    _____. (17.2)

16. Chromosomes contain DNA complexed with proteins called _____. (17.2)

17. During the S phase of the cell cycle, duplicated chromosomes are produced that consist of two identical _____ _____ joined by a _____. (17.2)

18. A _____ is the smallest identified functional unit of DNA. (17.2)

19. The first step of DNA replication is the _____ of the DNA helices. (17.2)

20. During DNA replication, new nucleotides are positioned and linked together by _____ _____ enzymes. (17.2)

21. DNA replication occurs at multiple _____ _____ along the length of mammalian chromosomes. (17.2)

22. Permanent changes that occur in a molecule of DNA are called _____. (17.2)

23. Errors in DNA replication are often repaired by DNA _____ _____. (17.2)

24. In the process of transcription, a base sequence called a _____ indicates where a gene begins. (17.2)

25. The enzyme RNA _____ positions nucleotides during transcription. (17.2)

26. A sequence of 3 bases that specify a particular amino acid is called a _____. (17.2)

27. The anticodon of a _____ molecule is complementary to a codon of a _____ molecule. (17.2)

Short Answer

28. Why does DNA replication in mammalian cells occur at multiple points along a chromosome at the same time?

29. List three things that can cause a mutation to occur in DNA.

30. How does a primary transcript differ from mRNA?

31. List the three types of RNA and the function of each type.

Labeling

For each process depicted in Figure 17.2, identify the process and where it occurs in the cell.

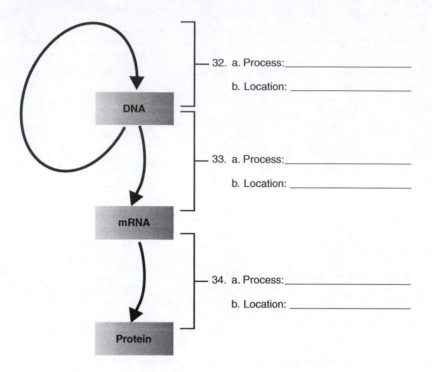

32. a. Process:_____

    b. Location: _____

33. a. Process:_____

    b. Location: _____

34. a. Process:_____

    b. Location: _____

Figure 17.2

Completion

DNA gene   **ACTACCCATGCCAG**

Use the base sequence of the DNA gene below to:

35. write the base sequence of the
    complementary DNA strand
    _____

36. write the base sequence of an mRNA
    strand produced from this gene
    _____

37. write the amino acid sequence of the
    protein produced from this gene,
    assuming this base sequence contains
    no introns
    _____

Ordering

Place the following events of translation in the appropriate column in the table below, and in the correct sequence.

a.   The newest amino acid is bonded to the last amino acid of the growing
     polypeptide chain.
b.   An initiator tRNA binds to the start codon of an mRNA.
c.   The large ribosomal subunit binds to the initiator tRNA and the small
     ribosomal subunit.
d.   The ribosome reaches a stop codon on the mRNA.

e. An initiator tRNA binds to the small ribosomal subunit.
f. A new tRNA molecule carrying the next amino acid binds to the ribosome and the mRNA.
g. The ribosomal subunits detach from the mRNA.
h. The RNA that delivered the newest amino acid is released from the ribosome.
i. The newly formed polypeptide chain is released from the ribosome.

Initiation	Elongation	Termination

17.3 Cell reproduction—one cell becomes two

17.4 How cell reproduction is regulated

17.5 Environmental factors influence cell differentiation

Define

Write the term(s) for each definition.

_____ 1. a phase of mitosis when duplicated chromosomes separate and move toward opposite sides of the cell

_____ 2. a cell that contains both members of each homologous pair

_____ 3. cells that have only one member of each homologous pair

_____ 4. a process in which duplicated homologous chromosomes pair up and exchange sections of DNA during prophase of meiosis I

_____ 5. an arrangement of microtubules that facilitates equal division of the DNA

_____ 6. chromosome pairs that have copies of the same genes in the same locations

_____ 7. phase of mitosis when duplicated chromosomes first become visible

_____ 8. a phase of mitosis when the two sets of chromosomes arrive at opposite poles of the cell

_____ 9. a phase of mitosis when duplicated chromosomes align at the center of the cell

Completion

Place an "X" in the column under mitosis, meiosis I, or meiosis II to indicate which cell division process includes the event described.

Event	Mitosis	Meiosis I	Meiosis II
10. crossing-over occurs between homologous chromosomes			
11. produces two haploid cells that still contain sister chromatids			
12. daughter cells are diploid			
13. in human cells 23 duplicated chromosomes will align on the equatorial plate			
14. daughter cells are exactly like the parent cell			
15. produces a secondary spermatocyte in a male			
16. produces an egg in a female			

Drawing

Complete each cell below by drawing in the chromosomes in their correct form and location. Begin with a diploid cell containing two pairs of chromosomes in prophase.

Prophase   Metaphase   Anaphase   Telophase and cytokinesis

Figure 17.3

Short Answer

A human liver cell is diploid and has 46 chromosomes. Answer the following questions about a liver cell.

17. How many chromosomes will be present in a liver cell in prophase?

18. How does the amount of DNA present in a prophase cell compare to the amount of DNA present in a cell in G1?

19. Does a cell in G2 contain duplicated chromosomes?

20. Is the DNA in a cell during G1 in the form of chromosomes or chromatin?

Critical Thinking

21. Study the diagram of meiosis in Figure 17.11 in your text. Meiosis occurs only in sexual reproduction, a process that supports survival of a species by ensuring genetic variation in the offspring. From the beginning of meiosis to fertilization, there are three events that provide opportunities for genetic variation in the offspring. What are these three events?

True/False

Mark each statement as true or false. If you mark a statement false, replace the underlined term or phrase with the term or phrase needed to make the statement true.

_____ 22. All cells in the body divide at <u>the same</u> rate.

_____ 23. Normal progression through the cell cycle is controlled by proteins called <u>cyclins</u>.

_____ 24. Checkpoints for progression through the cell cycle occur at the end of G1, the end of G2, and in <u>metaphase</u> of mitosis.

_____ 25. All of the cells in your body, with the exception of eggs and sperm, have the <u>same</u> DNA.

_____ 26. Cells in the body become specialized through the process of <u>differentiation</u>.

_____ 27. The physical and chemical environment of cells during development <u>does</u> effect how cells mature.

_____ 28. Cells develop along different lines due to <u>selective gene expression</u>.

17.6   Cloning an organism requires an undifferentiated cell

17.7   Therapeutic cloning: creating tissues and organs

Matching

Indicate whether each of the following statements describes a process or outcome of (a) embryo splitting or (b) somatic cell nuclear transfer by writing (a) and/or (b) beside each statement.

_____   1. produces a clone of an adult organism

_____   2. requires an enucleated fertilized egg

_____   3. requires a fertilized egg in the 8-cell stage

_____   4. a somatic cell is inserted into an enucleated fertilized egg

_____   5. produces multiple organisms that are identical to each other

_____   6. has been used to produce valuable farm animals

_____   7. a surrogate mother is required

_____   8. is a technique used for reproductive cloning

Looking Back/Looking Forward

1. Chapter 6 described muscle cells, and the structure of skeletal muscle cells that contain multiple nuclei. Considering the process of mitosis and cytokinesis, how might these multinucleate cells form?

2. Chapter 19 discusses genetics, and some of the inherited disorders that can result from errors during meiosis. One such disorder involves gametes that contain too many or too few chromosomes. What might happen in Anaphase I or Anaphase II that could produce such cells?

Chapter Test

Multiple Choice

1. Which phase of the cell cycle is incorrectly described?
   a. G_1—a time of slow cell growth
   b. S—sister chromatids are produced
   c. G_2—a cell prepares for cell division
   d. M—nuclear division occurs

2. Division of the cytoplasm occurs during
    a. mitosis and cytokinesis.
    b. interphase.
    c. DNA replication.
    d. cytokinesis.

3. Histone proteins
    a. contribute to the structure of a chromosome.
    b. become part of the mitotic spindle.
    c. disappear during prophase.
    d. direct chromosome movement during prophase.

4. A short segment of DNA that represents the code for a specific protein is a
    a. codon.
    b. triplet.
    c. gene.
    d. chromosome.

5. When a protein has been produced from the information in a molecule of DNA, which of the following has occurred?
    a. replication
    b. transcription
    c. translation
    d. b and c
    e. all of the above

6. The enzyme involved in positioning and joining complementary bases during replication is
    a. DNA replicase.
    b. DNA transcriptase.
    c. DNA polymerase.
    d. DNA nuclease.

7. A gene is expressed when
    a. DNA is replicated.
    b. mutations are repaired.
    c. a protein is produced.
    d. sister chromatids are produced.

8. Which of the following is a way in which replication and transcription differ?
    a. Only replication requires complementary base pairing.
    b. Only transcription requires an enzyme.
    c. Transcription copies only one side of the DNA molecule.
    d. Replication occurs in the nucleus.

9. A final mRNA transcript will contain
   a. introns and exons.
   b. introns only.
   c. exons only.
   d. a, b, or c, depending on the needs of the cell at the time.

10. _____ contains codons and _____ contains anticodons.
    a. tRNA, rRNA
    b. mRNA, tRNA
    c. mRNA, rRNA
    d. rRNA, tRNA

11. All proteins in the process of being produced will begin with the amino acid _____.

    a. Phe
    b. Ser
    c. Met
    d. Every protein begins with a different amino acid, depending on the DNA message.

12. In the process of translation, the instructions about the proper amino acid sequence of a new protein are carried by the
    a. mRNA.
    b. tRNA.
    c. rRNA.
    d. ribosome.

13. During the phase of translation called elongation, the mRNA passes between _____, catalyzing the binding of _____ to each other.
    a. rRNA and tRNA, proteins
    b. two ribosomal subunits, amino acids
    c. two strands of DNA, nucleotides
    d. a ribosome and tRNA, proteins

14. Which amino acid sequence is specified by the mRNA base sequence GAAUGCCCGCAC?
    a. glu-cys-pro-his
    b. leu-thr-gly-val
    c. met-pro-ala
    d. tyr-gly-arg

15. If the tubular elements of the cytoskeleton were not disassembled prior to mitosis, which of the following would be affected?
    a. formation of sister chromatids
    b. formation of the mitotic spindle
    c. disassembly of the nuclear membrane
    d. formation of a centromere

16. An original chromosome and its copy
    a. float singly in the cytoplasm during mitosis.
    b. are joined to each other by a centromere.
    c. are attached down the entire length of the chromosome.
    d. form a pair of homologous chromosomes.
    e. become sister chromatids.
    f. b and e

17. During which phase of mitosis do individual chromosomes arrive at opposite poles of the cell?
    a. metaphase
    b. anaphase
    c. prophase
    d. telophase

18. Which of the following would not occur if the cleavage furrow was unable to form?
    a. DNA replication
    b. division of the centromere
    c. cytokinesis
    d. formation of the mitotic spindle

19. Meiosis in a male produces _____ cells, each of which is _____.
    a. two, haploid
    b. two, diploid
    c. four, haploid
    d. four, diploid
    e. one, haploid

20. Genetic recombination in meiosis is accomplished by a process called
    a. mitosis.
    b. cleavage.
    c. homologous pairing.
    d. crossing-over.

21. A polar body is first produced
    a. during meiosis I in a female.
    b. during meiosis II in a female.
    c. during meiosis I in a male.
    d. during meiosis II in a male.

22. A mutation in the DNA of an organism is passed on to offspring only when
    a. it occurs during G_2 of the cell cycle.
    b. it occurs during the S phase of the cell cycle.
    c. it occurs in cells that give rise to an egg or a sperm.
    d. it affects both mitosis and cytokinesis.

23. Selective gene expression during development results in
    a. genetic mutations.
    b. more advanced single-celled organisms.
    c. the ability to clone an organism using cell separation.
    d. specialized cells.

24. Cells separated from each other after the 16-cell stage fail to produce an entire intact organism. This is because
    a. they require influential chemicals from neighboring cells in order to develop.
    b. the trauma of separating cells at this stage prevents further development.
    c. DNA at the 16-cell stage is difficult to activate.
    d. the cells have already begun to differentiate.

25. Most human cells contain _____ chromosomes, while the cells of gametes have _____ chromosomes.
    a. 23, 23
    b. 46, 46
    c. 23, 46
    d. 46, 23

Self Test

Each of the Key Concepts listed at the beginning of this chapter has been rewritten as a question below. After successfully completing the study guide exercises and the Chapter Test, you should be able to answer each of these questions. Refer to the Key Concepts list at the beginning of this chapter to check your answers.

1. What are the two primary phases of the cell cycle?

2. What three stages are included in interphase? What occurs during each stage?

3. What are the two divisions that make up the mitotic phase of the cell cycle?

4. How many chromosomes are found in human cells? What is a gene?

5. What is replication, and in what phase of the cell cycle does it occur? Describe the structure of a replicated chromosome.

6. What two processes must occur for the information in DNA to be used to make a protein?

7. What is transcription and where in the cell does it occur?

8. What is translation and where in the cell does it occur?

9. What are the three types of RNA that participate in translation, and what is the function of each?

10. What is a mutation? Are mutations always harmful? What process requires the presence of mutations?

11. What are the four phases of mitosis?

12. Describe the daughter cells produced by mitosis and meiosis as to the number of cells produced, the ploidy of the cells, and if they are genetically identical to the parent cell.

13. What are the two successive nuclear divisions of meiosis?

14. How is the product of meiosis different in males and females?

15. What factors help to control the cell cycle?

16. Name two things that lead to the development of specialized cells.

17. What is differentiation?

18. What is reproductive cloning, and by what two techniques is it accomplished?

19. How does the cloned product of embryo splitting differ from that of somatic cell nuclear transfer?

20. What is the goal of therapeutic cloning?

18

Cancer: Uncontrolled Cell Division and Differentiation

Chapter Summary and Key Concepts

After reading and studying this chapter, you should know the following:

Sections 18.1, 18.2, 18.3

1. Cancer is a disease of cell division and differentiation.

2. Normal cells have carefully regulated cell division, and they generally stay in one location in the body.

3. Cancer cells exhibit unregulated cell division, and may move to distant locations in the body.

4. A tumor is a mass of cells that has lost control over cell division. Tumors may be benign or cancerous.

5. Benign tumors remain in one place as a well-defined mass of cells.

6. A tumor is considered a cancer when the cells lose their organization, structure, and regulatory control.

7. Cancerous tumors that remain in one place are in situ cancers.

8. Malignant tumors are formed by cancer cells that invade normal tissues and metastasize to distant sites.

9. In order for cancer to develop, cells must divide uncontrollably and break away from surrounding cells.

10. Proto-oncogenes are regulatory genes that promote normal cell growth.

11. Tumor suppressor genes are normal regulatory genes that suppress cell growth.

12. Oncogenes are genes that cause cancer and develop from mutated proto-oncogenes.

13. Mutated or damaged proto-oncogenes and tumor suppressor genes contribute to the development of cancer by causing or allowing abnormal cell division.

14. Substances that cause cancer are called carcinogens, and they include viruses and bacteria, environmental chemicals, tobacco, radiation, dietary factors, and alcohol.

15. Susceptibility to cancer can be inherited if parents pass oncogenes to their offspring.

16. Suppression of the immune system may allow some cancers to develop more easily.

Sections 18.4, 18.5

17. Early detection is crucial in treating cancer successfully.

18. Traditional cancer treatments include surgery, radiation, chemotherapy, or a combination of two or more treatments.

Sections 18.6, 18.7

19. The 10 most common cancers are lymphomas, leukemia, and cancers of the skin, breast, prostate, lung, colon and rectum, urinary bladder, uterus, and kidney.

20. Most cancers can be prevented by making wise lifestyle and nutritional choices.

Exercises

Complete the exercises for each section after you have read and studied the section. If you cannot answer some questions, or if you answer them incorrectly, return to the chapter and review this information. You may find it helpful to work on only one section at a time. When you have completed all sections, take the Chapter Test as an indicator of your mastery of this topic.

18.1   Tumors can be benign or cancerous

18.2   Cancerous cells lose control over cell functions

18.3   How cancer develops

Define

Write the term(s) for each definition.

_____   1.   a substantial increase in the rate of cell division

_____   2.   a discrete mass formed by rapidly dividing cells

_____   3.   a type of tumor that remains in one place in a single well-defined mass

_____   4.   an abnormal structural change in a cell

_____   5.   the spread of cancer from its original site to another body region

_____   6.   genes that code for proteins required for cell growth and differentiation

_____   7.   genes that code for proteins that activate or suppress structural genes

_____   8.   proteins that influence nearby cells

_____   9.   a damaged proto-oncogene that contributes to cancer

_____   10.   a specific tumor suppressor gene known to inhibit cell division in cells with cancerous features

_____   11. a type of gene involved in DNA repair during replication that may contribute to cancer when damaged

_____   12. this is likely the single most important factor in the development of cancer

_____   13. the process of transforming a normal cell into a cancerous one

_____   14. a substance that causes cancer

Short Answer

15. What is indicated by hyperplasia?

16. When do benign tumors threaten health?

17. Describe the structural changes that occur in a cell when it becomes malignant.

18. When is a tumor defined as a cancer?

19. How does a malignant tumor differ from an in situ cancer?

20. What percentage of individuals in the United States will experience cancer in their lifetime? What percentage will die of cancer?

21. What two things must happen for cancer to develop?

22. What are the two classes of regulatory genes that control cell differentiation and division?

23. List the viruses believed to contribute to cancer.

24. What causes over 80% of all skin cancers?

25. What two factors are involved in approximately 30% of all cancers?

Labeling

Use the terms and descriptions to label Figure 18.1. Some structures will have two terms or descriptions that apply.

a. dysplasia   f. in situ cancer

b. malignant tumor   g. cells stay in one place

c. hyperplasia   h. cells change form

d. metastases   i. cancer cells invade normal tissue

e. genetically altered epithelial cell j. cells divide more rapidly than normal

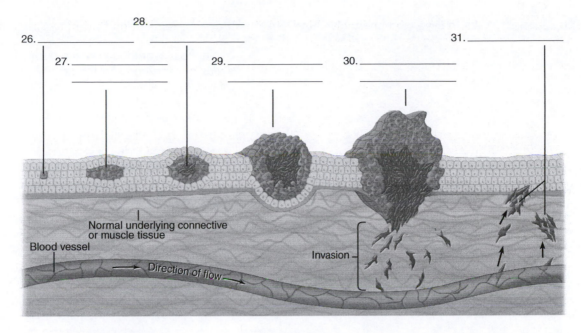

Figure 18.1

Critical Thinking

32. Women who have no or few children have an increased risk of ovarian cancer. What factors may contribute to the increased risk?

33. What characteristics of viruses may contribute to the viruses' ability to cause cancer?

18.4   Advances in diagnosis enable early detection

18.5   Cancer treatments

Completion

Name each diagnostic method and cancer treatment described in the table below.

Diagnostic Method or Treatment	Description
1.	a diagnostic method in which radioactive substances are used to create three-dimensional images showing the metabolic activity of body structures
2.	a diagnostic method which detects the presence of oncogenes

Diagnostic Method or Treatment	Description
3.	a diagnostic method in which body fluids are screened for the presence of telomerase
4.	a cancer treatment involving the administration of cytotoxic chemicals to destroy cancer cells
5.	a cancer treatment in which the cancer is targeted with lasers
6.	a potential cancer treatment in which defective genes are repaired or replaced
7.	a cancer treatment involving the use of anti-angiogenic drugs

Short Answer

8. Why are chemicals that inhibit cell division able to target primarily cancer cells while sparing most healthy cells?

9. How can the antigen of a cancer cell be used to develop immunotherapy treatments?

18.6 The 10 most common cancers

18.7 Most cancers can be prevented

Short Answer

Write the symptoms and risk factors for each cancer listed below. Answers for exercises in this section are not in the answer key; refer to the textbook to check your answers.

1. Skin cancer:

2. Lung cancer:

3. Breast cancer:

4. Prostate cancer

5. Colon and rectal cancer:

6. Lymphoma:

7. Urinary bladder cancer:

8. Kidney cancer:

9. Uterine cancer:

10. Cervical cancer:

11. Leukemia:

12. Using the recommendations in the textbook for cancer prevention, list at least six things you can do to reduce your risk of cancer:

    1. _____

    2. _____

    3. _____

    4. _____

    5. _____

    6. _____

Looking Back/Looking Forward

1. In what stage of the cell cycle are mutator genes active?

2. Chapter 19 describes the processes involved in inheritance. If your parents carry cancer-causing genes that you inherit, are you destined to develop cancer?

Chapter Test

Multiple Choice

1. Cell division in healthy cells
   a. is controlled by regulatory genes.
   b. is inhibited by signals from nearby cells.
   c. allows cells to travel through the body.
   d. a and b
   e. a, b, and c

2. Loss of regulatory control over cell division causes a cell to
   a. grow unusually large.
   b. secrete inhibiting factors.
   c. divide more frequently than normal.
   d. all of the above.

3. Benign tumors threaten health only if they
   a. secrete destructive enzymes.
   b. grow large enough to crowd normal cells.
   c. become cancerous.
   d. metastasize.

4. A tumor is defined as cancer when
   a. it metastasizes.
   b. the cells have lost regulatory control over cell division.
   c. the cells have undergone structural changes.
   d. some of the cells have lost regulatory control over cell division and undergone structural changes.

5. Select the sequence of developments that occurs in forming a malignant tumor, beginning with a single cell.
   a. single mutated cell, dysplasia, hyperplasia, malignant tumor, in situ cancer
   b. single mutated cell, hyperplasia, dysplasia, in situ cancer, malignant tumor
   c. single mutated cell, in situ cancer, dysplasia, hyperplasia, malignant tumor
   d. single mutated cell, dysplasia, in situ cancer, hyperplasia, malignant tumor

6. Which of the following is the same in an in situ tumor and a malignant tumor?
   a. cell structure
   b. cell function
   c. frequency of cell division
   d. tumor organization

7. Which of the following does not occur as a cell progresses toward becoming cancerous?
   a. The nucleus may enlarge.
   b. There may be less cytoplasm.
   c. Cell function is impaired.
   d. The cell retains its normal structures.

8. Cancer ranks _____ as a cause of death.
   a. first
   b. second
   c. third
   d. fourth

9. Tumor suppresor genes are
   a. normal regulatory genes that promote cell division.
   b. precancerous genes that have mutated at least once.
   c. normal regulatory genes that repress cell growth.
   d. regulatory genes that produce proteins that inhibit oncogenes.

10. Which of the following is *not* true of the gene p53?
    a. p53 encodes a gene that prevents damaged cells from dividing
    b. p53 is a tumor suppressor gene
    c. individuals who carry the p53 gene have an increased risk of cancer
    d. a mutated p53 gene has been found in some cervical cancers

11. A cell is dividing under circumstances when cell division would normally be inhibited. Which of the following is a regulatory gene that has probably been inactivated in the cell?
    a. a proto-oncogene
    b. an oncogene
    c. a tumor suppressor gene
    d. a growth factor

12. A cell has lost its ability to repair DNA errors that occur during DNA replication. What gene has probably been mutated?
    a. mutator gene
    b. oncogene
    c. proto-oncogene
    d. tumor suppressor gene

13. One mechanism by which viral infections may increase the risk of cancer is that
    a. viral infections increase the activity of tumor suppressor genes.
    b. the viral DNA sequence alters the function of a proto-oncogene.
    c. viral proteins interfere with maintenance of the plasma membrane.
    d. the virus interferes with the cell's ability to display "self" markers on the cell surface.

14. Sources of radiation that can increase cancer risk include
    a. the sun.
    b. medical and dental X-rays.
    c. radon gas.
    d. all of the above.

15. Diet and tobacco together account for approximately _____ of all cancers.
    a. 10%
    b. 40%
    c. 60%
    d. 90%

16. Inheriting a mutated proto-oncogene from one or both of your parents means that
    a. you will get cancer at a younger age than normal.
    b. you will get cancer, although it may be at an older age.
    c. you are at an increased risk for cancer.
    d. your risk for cancer is no different from anyone else's.

17. Which of the following statements about sunlight and skin cancer is *true*?
    a. Sunlamps and tanning booths are not as dangerous as natural sunlight.
    b. Sun exposure is only dangerous when it results in a sunburn.
    c. Any cumulative exposure to the sun increases the risk of cancer.
    d. Sunburns experienced during childhood do not increase the risk of cancer.

18. The presence of free radicals may increase the risk of cancer because
    a. free radicals reduce a cell's ability to fight viral infections.
    b. free radicals increase sun damage to cells.
    c. free radicals can damage a cell's DNA.
    d. free radicals interfere with normal DNA repair mechanisms.

19. Antioxidant activity
    a. inhibits the action of free radicals.
    b. is found in vitamin K.
    c. can reverse malignant changes in a cell.
    d. can convert a malignant tumor into a benign tumor.

20. In order for cancer to develop, what two things must happen?
    a. Both a tumor suppressor gene and an oncogene must be mutated.
    b. Cells must divide uncontrollably, and they must experience physical changes that allow them to break away from surrounding cells.
    c. Dietary and environmental factors that act as carcinogens must be present.
    d. A benign tumor must form, followed by hyperplasia.

21. An enzyme that occurs rarely in normal cells but frequently in cancer cells is
    a. prolactase.
    b. mutase.
    c. DNAse.
    d. telomerase.

22. The side effects of chemotherapy include nausea. This is because
    a.   the toxins in the drugs are difficult for the body to metabolize.
    b.   rapidly dividing healthy cells, such as occur in the lining of the digestive tract, are also destroyed.
    c.   chemotherapy interferes with the normal absorption of nutrients from the digestive tract.
    d.   as the chemicals are metabolized, the by-products increase acid production in the stomach.

23. The growth of new blood vessels to support rapidly growing tumors is called
    a.   angiogenesis.
    b.   cardiovascular supplementation.
    c.   vascularization.
    d.   capillary regeneration.

24. Which of the following skin cancers is incorrectly matched with its characteristic?
    a.   basal cell cancer—metastasizes quickly
    b.   squamous cell cancer—arises from epithelial cells
    c.   melanoma—the least common form of skin cancer
    d.   melanoma—the deadliest form of skin cancer

25. Cancer of the blood-forming organs
    a.   is called lymphoma.
    b.   occurs more often in children than adults.
    c.   is usually treated by radiation.
    d.   may be caused by a virus.

Self Test

Each of the Key Concepts listed at the beginning of this chapter has been rewritten as a question below. After successfully completing the study guide exercises and the Chapter Test, you should be able to answer each of these questions. Refer to the Key Concepts list at the beginning of this chapter to check your answers.

1. Cancer is a disease involving what two cellular processes?

2. What are the two primary characteristics of normal cells that are altered in cancer cells?

3. What are the two defining characteristics of cancer cells?

4. What characterizes the cell division of cells in a tumor? Are tumors always cancerous?

5. Describe a benign tumor.

6. What three things must tumor cells lose to be considered cancerous?

7. What is an in situ cancer tumor?

8. What is a malignant tumor?

9. What two things must happen in order for cancer to develop?

10. What are proto-oncogenes?

11. What are tumor suppressor genes?

12. What are oncogenes? How do oncogenes develop?

13. How do mutated proto-oncogenes or tumor suppressor genes contribute to the development of cancer?

14. What is a carcinogen? Name some common carcinogens.

15. How can susceptibility to cancer be inherited?

16. When does the immune system contribute to the development of cancer?

17. What is crucial in treating cancer successfully?

18. What are four traditional cancer treatments?

19. What are the 10 most common cancers?

20. Is cancer preventable?

19

Genetics and Inheritance

Chapter Summary and Key Concepts

After reading and studying this chapter, you should know the following:

Sections 19.1, 19.2, 19.3, 19.4

1. Genetics is the study of genes and their inheritance.

2. Human cells contain 22 pairs of chromosomes called autosomes and one pair of sex chromosomes.

3. Alleles are structural and functional variations of homologous genes.

4. An individual is homozygous when he or she possesses two identical alleles, and heterozygous when he or she possesses two different alleles.

5. An individual's complete set of genes and alleles is the genotype, and his or her observable physical and functional traits is the phenotype.

6. A Punnett square is a method for determining patterns of inheritance.

7. Gregor Mendel proposed the basic rules of inheritance in the 1850s.

8. The law of segregation states that gametes receive only one allele of each gene pair.

9. The law of independent assortment states that genes for different traits are separated from each other during meiosis.

10. In complete dominance, a dominant allele masks the effect of a recessive allele.

11. In incomplete dominance, the heterozygous genotype produces a phenotype that is intermediate between the two homozygous phenotypes.

12. In codominance, both alleles present will be equally expressed.

13. In polygenic inheritance, a phenotypic trait depends on the action of multiple genes.

14. Many phenotypes are determined by genotype and environmental effects.

15. Linked genes are genes located on the same chromosome and may be inherited together.

Sections 19.5, 19.6, 19.7, 19.8

16. A non-sex gene located on a sex chromosome shows a pattern of sex-linked inheritance.

17. A trait that is not carried on a sex chromosome but is influenced by the expression of a sex chromosome is a sex-influenced trait.

18. Chromosome structure or number may be altered by nondisjunction, deletions, and translocations.

19. Nondisjunction occurs when homologous chromosomes or sister chromatids fail to separate during cell division, a deletion occurs when a piece of a chromosome breaks off and is lost, and a translocation occurs when a piece of a chromosome breaks off and reattaches at another site.

20. Inherited genetic disorders usually involve recessive alleles.

Exercises

Complete the exercises for each section after you have read and studied the section. If you cannot answer some questions, or if you answer them incorrectly, return to the chapter and review this information. You may find it helpful to work on only one section at a time. When you have completed all sections, take the Chapter Test as an indicator of your mastery of this topic.

19.1   Your genotype is the genetic basis of your phenotype

19.2   Genetic inheritance follows certain patterns

19.3   Other dominance patterns

19.4   Other factors influence inheritance patterns and phenotype

Define

Write the term(s) for each definition.

_____   1. a DNA sequence that codes for one or more specific proteins

_____   2. all of the DNA belonging to an organism

_____   3. the study of genes and their transmission

_____   4. the chromosomes that do not determine gender

_____   5. a pair of chromosomes that have the same genes at the same locations on the chromosome

_____   6. the location of a gene on a chromosome

_____   7. alternative versions of a gene

_____   8. when a person has two identical alleles of a particular gene

_____   9. when a person has different alleles of a particular gene

_____   10. a random change to DNA that produces alleles

_____ 11. an individual's complete set of alleles

_____ 12. an individual's observable physical and functional traits

Short Answer

13. How many pairs of autosomes do humans have in each somatic cell? How many pairs of sex chromosomes?

14. If there are more than two alleles present in a population for a particular trait, how many will an individual carry?

15. What is stated in the Law of Segregation? In the Law of Independent Assortment? Can the Law of Independent Assortment be observed in a monohybrid cross?

16. How is a dominant allele represented? A recessive allele? When is a recessive trait observed?

17. How does codominant inheritance differ from incomplete dominance?

18. List some traits that result from polygenic inheritance.

19. When are linked genes most likely to be inherited together? What process can sometimes separate them?

Labeling

Write in the genotypes of the male and female parent in Figure 19.1, and then complete the Punnett square.

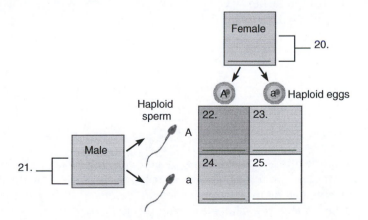

Figure 19.1

Short Answer

26. Albinism, a trait in which individuals have no pigment, is inherited as a recessive trait (a), while normal pigment is carried by the dominant allele (A). Only individuals who are homozygous will be albino.

    a. What is the genotype of an albino individual?

    b. What is the phenotype of an individual whose genotype is Aa?

    c. What is the genotype of an individual who is heterozygous?

    d. What is the phenotype of an individual who is homozygous dominant?

    e. Use a Punnett square to determine the possible genotypes of all children born to a heterozygous father and an albino mother. What percentage of the children will be expected to be albino?

27. In the inheritance of hair type, homozygous dominant (WW) individuals have curly hair, homozygous recessive (ww) individuals have straight hair, and heterozygous (Ww) individuals have wavy hair.

    a. What type of inheritance is occurring?

    b. Use a Punnett square to determine the genotypes of the children born to a curly-haired father and a straight-haired mother.

    c. Use a Punnett square to determine the genotypes of the children born to an albino father with curly hair and a heterozygous normally pigmented mother with wavy hair.

28. The shape of the hemoglobin molecule in a red blood cell is a genetic trait. The normal hemoglobin gene (Hb^A) codes for a normal-shaped red blood cell, while the sickle-cell gene (Hb^S) codes for an abnormally shaped hemoglobin molecule that becomes sickle-shaped during physiological stress. Sickle-cell is inherited in a codominant fashion, although the phenotype appears more like an incompletely dominant trait. Individuals who carry two normal genes have normal red blood cells, individuals who carry one of each have sickle-cell trait with some hemoglobin molecules of each type, and individuals who carry two sickle cell genes have sickle-cell anemia and no normal hemoglobin molecules.

a.   What is the phenotype of an individual who is heterozygous?

b.   What is the genotype of an individual with sickle-cell anemia?

c.   Use a Punnett square to determine the possible genotypes of all children born to a father with Sickle-cell trait, and a mother who has sickle-cell anemia. What percentage of the children can be expected to have only normal hemoglobin molecules? Why is this a codominant trait, and why does the phenotype appear more like a trait that is incompletely dominant?

Critical Thinking

29. Your friend is pregnant and has been recently diagnosed with skin cancer. She is very concerned that her child will inherit a predisposition for skin cancer and asks you to explain the degree of risk involved. How would you address her concern?

19.5   Sex-linked inheritance: X and Y chromosomes carry different genes

19.6   Chromosomes may be altered in number or structure

19.7   Many inherited genetic disorders involve recessive alleles

19.8   Genes code for proteins, not for specific behaviors

Fill-in-the-Blank

Referenced sections are in parentheses.

1.   A composite display of all of the chromosomes of an organism is called a

_____. (19.3)

2.   Human cells contain 22 pairs of chromosomes called _____ and one pair of

_____ chromosomes. (19.3)

3.   In humans, maleness is determined by the presence of a _____

chromosome. (19.3)

4.   An inheritance pattern that depends on genes located on sex chromosomes is called

_____ _____ inheritance. (19.3)

5. A trait that is not inherited on a sex chromosome but is influenced by the expression of a sex chromosome is called a _____ _____ trait. (19.3)

6. _____ occurs when homologous chromosomes or sister chromatids fail to separate during meiosis. (19.4)

7. A _____ occurs when a piece of a chromosome breaks off and is lost. (19.4)

8. A _____ occurs when a piece of a chromosome breaks off and reattaches at another site. (19.4)

9. The goal of the _____ _____ project is to identify the location and function of all genes in the human genome. (19.6)

Short Answer

10. Why do more males than females have X-linked disorders?

11. Why can a man inherit an X-linked trait only from his mother?

12. Why will all the daughters of a man with an X-linked trait be carriers for the disorder?

13. Hemophilia is inherited as an X-linked recessive trait. Use X^H to represent the allele for normal blood, and X^h to represent the allele for hemophilia.
    a. What is the genotype of a man who has hemophilia?
    b. What is the genotype of a woman who has hemophilia?
    c. What is the genotype of a man who has normal blood?
    d. What is the genotype of a woman who has normal blood?
    e. What is the genotype of a woman who has normal blood and is a carrier of hemophilia?
    f. Use a Punnett square to determine the possible genotypes of children born to a man with hemophilia and a woman who has normal blood but is a carrier of the trait. What percentage of their sons can be expected to have normal blood? What percentage of their daughters can be expected to have hemophilia?

14. Why are sex-influenced traits, such as pattern baldness, expressed differently in men and women who both have the heterozygous genotype?

15. Why might the ability of genes to encode proteins influence behavior and personality?

Completion

Identify each genetic disorder described in the table below.

Genetic Disorder	Description
16.	Individuals with this disorder have three X chromosomes, are normal females, and often have mild retardation.
17.	This disorder is inherited as autosomal recessive and develops because affected individuals are unable to metabolize phenylalanine.
18.	This disorder is caused by a deletion in chromosome 5, and affected infants are usually mentally and physically retarded, with a cat-like cry.
19.	Affected men have the genotype XXY, are tall and sterile, and may have mild mental impairment.
20.	This disorder is most often caused by an extra copy of chromosome 21.
21.	This disorder is caused by a dominant-lethal allele and results in progressive nerve degeneration.
22.	Individuals with this disorder are male, tend to be tall, appear normal, and may have impaired mental function.

Labeling

Draw the chromosome in the cells below, indicating the results of nondisjunction at different stages of meiosis. (Refer to Figure 19.17 of the textbook to check your answers.)

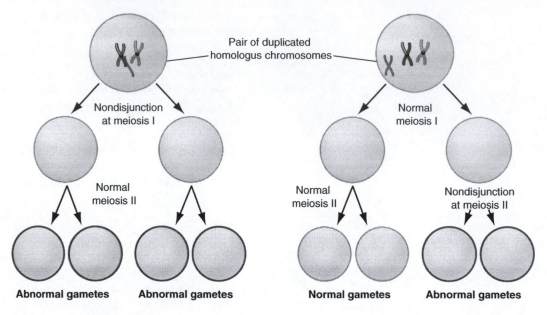

Figure 19.2

Looking Back/Looking Forward

1. Chapter 17 described the process of gene expression. Recessive alleles may have originated as mutations. Do recessive alleles code for the production of a functional protein?

2. Chapter 22 discusses evolution, including human evolution, and the processes that lead to evolution. Considering what you have learned about mutations and inheritance, do you think humans are currently evolving? Explain your answer.

Chapter Test

Multiple Choice

1. Human cells contain 22 pairs of _____ .
   a. chromosomes
   b. alleles
   c. homologues
   d. autosomes

2. An individual whose genotype is Ww is referred to as being
   a. homozygous.
   b. heterozygous.
   c. genotypic.
   d. dominant.

3. The law of segregation states that
   a. in the formation of gametes, alleles separate from each other so that each gamete receives only one allele.
   b. male and female gametes segregate independently of each other.
   c. the genes on one chromosome move independently of the genes on another chromosome.
   d. only alleles that are not linked will be segregated during the formation of gametes.

4. A recessive phenotype will be expressed when
   a. at least one recessive allele is present.
   b. at least one dominant allele is present.
   c. both alleles present are recessive.
   d. both alleles present are dominant.

5. In complete dominance, the recessive allele usually
   a. codes for only a small amount of a protein.
   b. codes for a fully functional protein.
   c. codes for a nonfunctional protein or produces no protein.
   d. codes for a protein that is repressed by the dominant allele.

6. The law of independent assortment can be observed
   a. in a one-trait cross.
   b. in a two-trait cross.
   c. in both one-trait and two-trait crosses.
   d. only during mitosis.

7. In the human ABO blood-typing system, an individual with A and B alleles will express both A and B phenotypes on their red blood cells. This is an example of
   a. complete dominance.
   b. incomplete dominance.

    c.   codominance.

    d.   polygenic inheritance.

8. In humans, the allele for a widow's peak (W) is dominant. Which of the following represent the possible phenotypes in the offspring of two parents, both with a straight hairline?

    a.   All offspring will have a widow's peak.

    b.   All offspring will have a straight hairline.

    c.   Offspring will show both phenotypes.

    d.   Most of the offspring will have a straight hairline, but it is possible to have a widow's peak.

9. When continuous variation occurs in a phenotypic trait, the trait is likely due to

    a.   complete dominance.

    b.   incomplete dominance.

    c.   codominance.

    d.   polygenic inheritance.

10. The frequency with which genes on a particular chromosome are inherited together depends on

    a.   whether they are dominant or recessive.

    b.   the occurrence of nondisjunction during meiosis.

    c.   how close together they are on the chromosome.

    d.   how often crossing over occurs.

    e.   c and d

11. In humans, femaleness is determined by

    a.   the presence of at least one X chromosome.

    b.   the presence of two X chromosomes.

    c.   the absence of a Y chromosome.

    d.   the presence of adequate levels of estrogen.

12. Testosterone is responsible for the development of the male reproductive system, and it is produced by the testis. The development of the testes depends on the presence of a Y chromosome. What might occur if the Y chromosome is present, but the testes are unable to produce testosterone?

    a.   normal male development

    b.   normal female development

    c.   the testes would be present, but further male development would be arrested

    d.   testes and ovaries would be present

13. Hemophilia is an X-linked recessive trait resulting in the inability to clot the blood. A man with hemophilia inherited the condition from

    a.   his mother.

    b.   his father.

    c.   both of his parents.

    d.   all of the above could produce hemophilia in a man.

14. A woman with hemophilia inherited the condition from
    a. her mother.
    b. her father.
    c. both of her parents.
    d. all of the above could produce hemophilia in a woman.

15. Baldness is a sex-influenced trait. Women must inherit both recessive alleles to be bald, while men need only inherit one recessive allele. This occurs because
    a. estrogen masks the effect of the baldness allele.
    b. the presence of two X chromosomes in a woman masks the baldness allele.
    c. the baldness allele is linked to the Y chromosome.
    d. testosterone stimulates the expression of the baldness allele.

16. Nondisjunction in meiosis I results in failure of
    a. homologous chromosomes to separate.
    b. sister chromatids to separate.
    c. gametes to separate.
    d. centromeres to separate.

17. An infant born with a deletion in chromosome 5 will have
    a. Down syndrome.
    b. Turner syndrome.
    c. Edwards' syndrome.
    d. cri du chat syndrome.

18. An infant born with the genotype XYY will
    a. be phenotypically female and fertile.
    b. be phenotypically male, short, and mildly retarded.
    c. be phenotypically male and usually normal.
    d. show partial male and female development.

19. Dominant alleles that cause disease are especially dangerous because
    a. the severity of the disease is greater than those caused by recessive alleles.
    b. heterozygous individuals will develop the disease.
    c. individuals must inherit two dominant alleles to show the disease.
    d. dominant alleles are inherited more frequently than recessive alleles.

20. Which of the following is not a source of genetic variability during sexual reproduction?
    a. independent assortment
    b. crossing-over between homologous chromosomes
    c. polygenic inheritance
    d. random fertilization of an egg by a sperm

21. Alterations in chromosomal structure or number that occur during mitosis or meiosis are seldom seen because
    a. they do not affect the organism.
    b. the effects are so mild that they often go unrecognized.
    c. the effects are usually masked by more serious conditions.
    d. the altered genes are crucial to development and the organism dies.

22. A man carries a Y-linked trait. He will pass this trait to
    a. all of his daughters.
    b. all of his sons.
    c. 50% of his daughters and 100% of his sons.
    d. 50% of his sons and none of his daughters.

23. When a translocation occurs
    a. a piece of a chromosome breaks off and is lost.
    b. a damaged chromosome is passed from parent to child.
    c. an X-linked trait is passed to a son.
    d. a segment of a chromosome breaks off and reattaches at another location.

24. Which of the following is a condition caused by a dominant-lethal allele?
    a. Down syndrome
    b. Huntington's disease
    c. Klinefelter syndrome
    d. Hemophilia

25. Specific genes code for specific
    a. proteins.
    b. behaviors.
    c. traits.
    d. allele combinations.

Self Test

Each of the Key Concepts listed at the beginning of this chapter has been rewritten as a question below. After successfully completing the study guide exercises and the Chapter Test, you should be able to answer each of these questions. Refer to the Key Concepts list at the beginning of this chapter to check your answers.

1. What is genetics?

2. How many pairs of chromosomes, and of what type, do human cells contain?

3. What are alleles?

4. What makes an individual homozygous or heterozygous?

5. How would you define genotype and phenotype?

6. What is a Punnett square used for?

7. Who proposed the basic rules of inheritance, and in what year?

8. What principle is stated by the law of segregation?

9. What principle is stated by the law of independent assortment?

10. What occurs in complete dominance?

11. How is incomplete dominance different from complete dominance?

12. How are alleles expressed in codominance?

13. What determines phenotype in polygenic inheritance?

14. What, besides genotype, can influence the phenotypic expression of a trait?

15. What are linked genes?

16. What causes sex-linked inheritance?

17. What is a sex-influenced trait?

18. What are three ways in which chromosome number or structure can be altered?

19. What occurs in nondisjunction? What is a deletion? What is a translocation?

20. Do inherited disorders usually involve recessive or dominant alleles?

20

DNA Technology and Genetic Engineering

Chapter Summary and Key Concepts

After reading and studying this chapter, you should know the following:

Sections 20.1, 20.2

1. Biotechnology is the technical application of biological knowledge for human purposes.

2. DNA sequencing allows scientists to determine the base sequence of a DNA molecule.

3. DNA sequencing relies on complementary base pairing to produce a new DNA strand, and requires the use of short single-stranded DNA segments, primers, modified DNA nucleotides, DNA polymerase, gel electrophoresis, and a fluoroscope.

4. Recombinant DNA is produced by combining DNA from two different sources. Recombinant DNA technology provides tools for manipulating DNA, and involves cutting, splicing, and copying DNA.

5. Recombinant bacteria are produced using DNA from a foreign organism, restriction enzymes, DNA ligases, and plasmids.

6. Restriction enzymes cut DNA by breaking the bonds between specific base pairs.

7. DNA ligases can bind DNA fragments from different organisms together after they have been cut by restriction enzymes.

8. Plasmids are extrachromosomal DNA molecules that occur naturally in bacteria.

9. Plasmids can accept a foreign piece of DNA and carry it back into the bacteria. Bacteria containing a foreign gene can manufacture a foreign protein.

10. Polymerase chain reaction (PCR) is a technique that amplifies a small fragment of DNA, using separated strands to serve as templates for the construction of new complementary strands. Cycles of heating and cooling allow the process to repeat until many identical DNA molecules are produced.

11. DNA fingerprinting identifies the source of a DNA fragment after it has been amplified, using characteristics of non-coding DNA sequences within the fragment.

Sections 20.3, 20.4

12. Transgenic organisms carry foreign genes as a result of genetic engineering.

13. Transgenic bacteria created with engineered plasmids can produce foreign proteins, including human proteins.

14. Transgenic plants carry foreign genes and have been engineered to resist adverse weather and pests, carry vitamin precursors, and produce vaccines.

15. Transgenic plants have created some concerns about potential safety and environmental hazards associated with their use.

16. Creating transgenic animals is difficult, but research continues because of the great potential benefits.

17. Current techniques for creating transgenic animals include inserting DNA into a single isolated cell, which can give rise to a whole organism.

18. Gene therapy involves the insertion of human genes into human cells for medical purposes.

19. Gene therapy may utilize retroviruses as vectors—agents that can transport a gene into a human cell.

20. Gene therapy has been successfully used to treat severe combined immune deficiency, and research continues in the treatment of cystic fibrosis and cancer.

Exercises

Complete the exercises for each section after you have read and studied the section. If you cannot answer some questions, or if you answer them incorrectly, return to the chapter and review this information. You may find it helpful to work on only one section at a time. When you have completed all sections, take the Chapter Test as an indicator of your mastery of this topic.

20.1   DNA sequencing reveals structure of DNA

20.2   DNA can be modified in the laboratory

Define

Write the term(s) for each definition.

_____   1. the technical application of biological knowledge for human purposes

_____   2. manipulation of the genetic makeup of cells or organisms

_____   3. the enzyme that adds new DNA nucleotides to a growing strand of DNA

_____   4. enzymes that break the bonds between specific base pairs in a DNA strand

_____   5. a DNA nucleotide sequence that reads the same from left to right on one strand as it does from right to left on the complementary strand

_____   6. enzymes that bind fragments of DNA together

_____   7. small, circular, self-replicating DNA molecules found in bacteria

_____   8. a technique used to amplify a small amount of DNA

_____   9. a technique that identifies the source of a DNA fragment after it has been amplified

True/False

The steps of DNA sequencing are listed below in their correct sequence. Mark each statement as true or false. If you mark a statement false, replace the underlined term or phrase with the term or phrase needed to make the statement true.

_____   10. Millions of identical copies of a short <u>single-stranded</u> piece of DNA are placed in a test tube.

_____   11. <u>Primers</u> are added to serve as the beginning point for synthesis of each new complementary DNA strand.

_____   12. A primer is a short segment of <u>double-stranded</u> DNA.

_____   13. A, T, C, and G <u>nucleotides</u> are added to the test tube.

_____   14. Additional nucleotides are added that have been modified, as well as being tagged with fluorescent markers for <u>identification</u>.

_____   15. DNA <u>ligase</u> is added to the test tube.

_____   16. Synthesis of a new complementary DNA strand <u>continues</u> when a modified nucleotide is added.

_____   17. As synthesis of new DNA strands occurs, many pieces of double-stranded DNA will be produced at <u>various</u> lengths.

_____   18. Strands of DNA produced in the test tube will each end with a labeled and identifiable <u>modified</u> nucleotide.

_____   19. DNA strands are separated by <u>gel electrophoresis</u> according to their length.

_____   20. An electrical field causes the DNA fragments to migrate through the gel, with smaller fragments moving <u>faster</u> than larger fragments.

_____   21. A <u>laser</u> detects the four different fluorescent labels as the DNA fragments come off the gel.

_____   22. As the fragments are arranged by length, the labeled nucleotides at the end of each fragment produce the base sequence of the <u>original fragment</u> of DNA.

Short Answer

23. How long have humans been involved in recombining DNA?

24. What is the function of restriction enzymes in nature?

25. What is the role of plasmids in nature? How are they used in recombinant DNA technology?

26. Why is PCR not a useful technique for cloning whole genes and their protein products?

27. What two aspects of DNA make DNA fingerprinting useful in identifying the source of a DNA fragment?

Sequencing

Place the following events in the production of recombinant DNA in their proper sequence by numbering them 1 through 6.

_____ 28. introduce the plasmid containing human DNA into bacteria

_____ 29. mix human DNA fragments with cut plasmids

_____ 30. select the bacteria containing the human DNA and allow them to reproduce

_____ 31. cut plasmids and human DNA with the same restriction enzymes

_____ 32. isolate plasmid and human DNA of interest

_____ 33. add DNA ligase to cut plasmids and human DNA fragments

20.3   Genetic engineering creates transgenic organisms

20.4   Gene therapy: The hope of the future

Crossword Puzzle

Across

6. Microinjections of DNA into a cell is successful less than _____% of the time.
7. Human insulin is currently produced in genetically engineered _____.
11. Cancer research includes the possibility of adding genes for _____ to a patient's cancer cells.
12. Gene therapy is one of the potential benefits of the Human _____ Project.
14. Transgenic animals are _____ to produce.
16. When transgenic plants incorporate human genes they produce human _____.
17. Gene therapy for cystic fibrosis may someday involve the use of _____ spray.
18. The process of producing pharmaceuticals in farm animals is called gene _____.
19. Transgenic bacteria aid the production of _____ when they produce foreign antigens.

Down

1. The preferred vector for creating transgenic animals.
2. A genetically engineered vaccine against _____ is currently on the market.
3. The creation of transgenic animals may begin with insertion of foreign DNA into _____ eggs.
4. Viruses used as vectors to create transgenic animals must first be genetically _____.
5. Concerns exist that the use of genetically engineered food crops may lead to some crop _____.
8. _____ organisms carry one or more foreign genes.
9. Something that transports genes into a foreign cell.
10. Gene therapy has successfully treated severe combined _____ deficiency disease.
13. Fatty sphere, which may be able to carry recombinant plasmids into animal cells.
15. Gene _____ involves repairing or replacing damaged human genes.

Short Answer

20. List three human hormones that are currently being produced by transgenic bacteria.

21. Why are vaccines produced by transgenic bacteria safer than viruses made from weakened or killed disease-causing organisms?

22. What are two of the biggest challenges in gene therapy?

Critical Thinking

23. Cystic fibrosis is an inherited disorder in which the gene coding for a crucial transmembrane protein involved in the transport of chloride ions is affected. Early efforts to treat the disorder with gene therapy have been unsuccessful, but encouraging. Assuming successful gene therapy treatments are developed in the future, will affected individuals be able to avoid passing this disorder to their offspring? Explain your answer.

Looking Back/Looking Forward

1. Chapter 18 describes how viruses may cause cancer. Theoretically, could the use of viruses as vectors for gene transfer contribute to the development of cancer?

2. Chapter 24 looks at how human activity is altering our environment, often in damaging ways. List some ways transgenic bacteria can be used to reduce such damage.

Chapter Test

Multiple Choice

1. Which of the following functions as a primer in DNA sequencing?
   a. DNA polymerase
   b. nucleotides labeled with a fluorescent marker
   c. short, single-stranded pieces of DNA that bind to the end of a single-stranded piece of DNA
   d. short, double-stranded pieces of DNA that bind to the end of a single-stranded piece of DNA

2. Gel electrophoresis uses _____ to separate DNA pieces according to their _____.
   a. enzymes, bond structure
   b. primers, base sequence
   c. an electrical field, size
   d. a fluoroscope, altered DNA nucleotides

3. All of the following are associated with DNA sequencing except
   a. DNA nucleotides labeled with a fluorescent marker.
   b. single-stranded DNA primers.
   c. gel electrophoresis.
   d. plasmids incubated with DNA fragments.

4. Recombinant DNA
   a. has existed for as long as organisms have carried out sexual reproduction.
   b. results from the selective breeding of plants and animals.
   c. produced from the intentional cutting and splicing of DNA molecules has existed for only a few decades.
   d. all of the above.

5. In nature, plasmids
   a. are found in bacteria and help to protect the bacteria from viral infection.
   b. are found in viruses and help viral DNA splice itself into a host chromosome.
   c. are found in plants and carry specific genes that are used in plant reproduction.
   d. are found in animals and help to provide protection against bacterial infection.

6. Restriction enzymes cut DNA
   a. between specific base pairs.
   b. in palindromic sequences.
   c. between hydrogen-bonded base pairs.
   d. a and b
   e. a, b, and c

7. Cut fragments of DNA can be bound together by
   a. DNA polymerase.
   b. DNA ligase.
   c. plasmids.
   d. restriction enzymes.

8. Which of the following is(are) *not* required to create a recombinant plasmid?
   a. DNA polymerase
   b. DNA ligase
   c. foreign DNA
   d. restriction enzymes

9. Place the following steps in the order they would occur during the creation of recombinant plasmids.
   I.   addition of DNA ligase
   II.   isolate DNA from a foreign source
   III.   mix foreign DNA fragments with cut plasmids
   IV.   introduce recombinant plasmid into bacteria
   V.   cut foreign DNA and plasmids with restriction enzymes
   a. I, II, III, IV, V
   b. IV, I, II, V, III
   c. II, V, III, I, IV
   d. II, III, V, I, IV

10. When recombinant plasmids are created to clone human genes,
    a. retroviruses are required to complete the cloning process.
    b. bacteria containing the desired gene must be identified and isolated prior to cloning.
    c. the plasmids are able to reproduce and clone the genes independently, without bacteria.
    d. the gene of interest must first be microinjected into the bacteria.

11. The polymerase chain reaction (PCR) makes it possible to
    a. express all the genes in a particular DNA sequence.
    b. amplify a small amount of DNA.
    c. create a large quantity of DNA polymerase for use in genetic engineering.
    d. create transgenic organisms.

12. During PCR, DNA strands are unwound during _____, and complementary strands are formed during _____.
    a. heating, heating
    b. heating, cooling
    c. cooling, heating
    d. cooling, cooling

13. Transgenic organisms are organisms that
    a. are produced by selective breeding.
    b. carry recombinant plasmids.
    c. are resistant to diseases.
    d. carry one or more foreign genes.

14. Transgenic bacteria are currently used to produce all of the following except
    a. beta-carotene.
    b. insulin.
    c. human growth hormone.
    d. erythropoietin.

15. Bacteria that produce vaccines have been engineered to express the gene for
    a. specific antibodies.
    b. human proteins affected by the disease.
    c. the immune cells normally involved in fighting the disease.
    d. the antigen of the disease-causing organism.

16. Using genetically engineered bacteria to produce vaccines is difficult because
    a. a specific gene must be located and transferred to the bacteria.
    b. the process is time-consuming and expensive.
    c. microorganisms evolve too rapidly to guarantee that a vaccine will always be effective.
    d. all of the above.

17. Transgenic plants
    a. can be engineered to express bacterial genes, but not human genes.
    b. have been created, but useful applications have not yet been identified.
    c. do not pose any safety or environmental concerns.
    d. have been engineered for herbicide resistance and vitamin delivery.

18. New cancer research involves introduction, by way of a viral vector, of the mda-7 gene into malignant tumors. As a result of this therapy, malignant tumor cells
    a. begin to secrete cell-surface proteins that help the immune system to recognize the malignant cells.
    b. are unable to stimulate the growth of the new blood vessels required to nourish the tumor.
    c. lose the ability to metastasize.
    d. die as a result of apoptosis.

19. Transgenic animals are most often created when
    a. foreign DNA is microinjected into a fertilized egg.
    b. animal cells readily take up recombinant plasmids.
    c. animals are infected by engineered bacteria.
    d. a single defective gene is replaced by a normal gene.

20. Genetic engineering
    a. is widely accepted as a valuable tool.
    b. has the potential to provide cures for many human health concerns.
    c. raises many ethical problems.
    d. all of the above.

21. The ability to repair or replace damaged human genes is an example of
    a. gene therapy.
    b. recombinant DNA technology.
    c. genetic engineering.
    d. all of the above.

22. Which of the following is a *false* statement about gene therapy?
    a. Gene therapy has the potential to help in the treatment of many diseases.
    b. One of the challenges of gene therapy is finding a way to deliver new DNA to all necessary body cells.
    c. Diseases that are successfully treated with gene therapy cannot be inherited by an affected individual's offspring.
    d. Retroviruses are potential vectors for use in gene therapy.

23. The first disease to be treated successfully with gene therapy was
    a. cystic fibrosis.
    b. SCID.
    c. cancer.
    d. diabetes.

24. Retroviruses have the potential to be effective vectors in delivering DNA to human cells because
    a. they do not normally infect human cells.
    b. they never mutate.
    c. they are easily microinjected.
    d. they can splice their genetic information into the DNA of human cells.

25. While retroviruses are capable of introducing foreign DNA into a human genome,
    a. this has never been done successfully.
    b. they may disrupt important regulatory genes when they insert foreign DNA at random locations in a chromosome.
    c. their use is limited because they only insert foreign DNA during G1 of the host cell cycle.
    d. the difficulty of creating a recombinant virus makes their common use as a vector unlikely.

Self Test

Each of the Key Concepts listed at the beginning of the study guide has been rewritten as a question below. After successfully completing the study guide exercises and the Chapter Test, you should be able to answer each of these questions. Refer to the Key Concept List at the beginning of this chapter to check your answers.

1. What is biotechnology?

2. What does DNA sequencing allow scientists to determine?

3. What structures, enzymes, and processes are required for gene sequencing?

4. What is recombinant DNA? What three processes are involved in recombinant DNA technology?

5. List four things needed to produce recombinant bacteria.

6. What effect do restriction enzymes have on DNA?

7. What is the action of DNA ligase?

8. What are plasmids and where are they found in nature?

9. What can plasmids carry into bacteria? What must bacteria contain to manufacture a foreign protein?

10. Define PCR and briefly describe its process.

11. What is DNA fingerprinting?

12. What are transgenic organisms?

13. What are transgenic bacteria and what can they produce?

14. What are transgenic plants? List some examples of the benefits of transgenic plants.

15. What are some of the concerns associated with the use of transgenic plants?

16. Why does research on transgenic animals continue even though they are difficult to create?

17. What is one technique used to create transgenic animals?

18. What is gene therapy?

19. How are retroviruses used in gene therapy?

20. What disease has been successfully treated with gene therapy? List two diseases that are currently being investigated for treatment with gene therapy.

21

Development and Aging

Chapter Summary and Key Concepts

After reading and studying this chapter, you should know the following:

Sections 21.1, 21.2

1. Sperm travel through the vagina, cervix, uterus, and oviduct in an attempt to fertilize the egg.

2. The egg is released as an immature secondary oocyte and completes meiosis II only after fertilization occurs.

3. Fertilization occurs in the upper third of the oviduct when the nuclei of the sperm and ovum fuse to form a zygote.

4. Twins may be fraternal, identical, or conjoined.

5. The four processes of development are: a series of cell divisions called *cleavage,* a process of physical change called *morphogenesis,* the development of specialized form and function in cells called *differentiation,* and an increase in size called *growth.*

Sections 21.3, 21.4

6. Prenatal development is divided into three stages: pre-embryonic, embryonic, and fetal development.

7. In pre-embryonic development, the zygote becomes a morula and then a blastocyst that implants in the endometrium.

8. During embryonic development the embryo consists of three germ layers called the endoderm, the mesoderm, and the ectoderm.

9. Four extra-embryonic membranes form during embryonic development to support the embryo: the amnion, the allantois, the yolk sac, and the chorion.

10. The placenta consists of a maternal portion, the endometrium, and an embryonic portion, the chorion and chorionic villi.

11. The umbilical cord connects the placenta to the embryo's circulation.

12. By the end of embryonic development the embryo is distinctly human in appearance.

Sections 21.5, 21.6, 21.7

13. During fetal development, growth is rapid, gender development begins, the fetus begins to move, surfactant is produced, and life outside the womb becomes possible.

14. Birth involves three stages of labor and delivery: dilation, expulsion, and afterbirth.

15. At birth, changes in the fetal circulation sustain breathing by the newborn.

16. Lactation delivers first colostrum, and then milk, to the newborn.

Sections 21.8, 21.9, 21.10

17. The stages of human development after birth are the neonatal period, infancy, childhood, adolescence, and adulthood.

18. Aging is a process of deterioration that occurs over time.

19. Aging is marked by changes in major body systems.

20. Death is the irreversible cessation of processes required to sustain life.

Exercises

Complete the exercises for each section after you have read and studied the section. If you cannot answer some questions, or if you answer them incorrectly, return to the chapter and review this information. You may find it helpful to work on only one section at a time. When you have completed all sections, take the Chapter Test as an indicator of your mastery of this topic.

21.1   Fertilization occurs when sperm and egg unite

21.2   Development: Cleavage, morphogenesis, differentiation, and growth

Labeling and Short Answer

Use the terms below to label Figure 21.1 of a sperm and a secondary oocyte, and then answer the questions that follow.

head	**nucleus**	**corona radiata**	**cytoplasm**
midpiece	**first polar body**	**acrosome**	**plasma membrane**
zona pellucida	**tail**		

1. _____

2. _____

3. _____

Oviduct

Uterus

5. _____

Secondary oocyte

6. _____

Sperm

Ovary

7. _____

Cervix

8. _____

Vagina

9. _____

4. _____

10. _____

Figure 21.1

11. What part of a sperm contains enzymes that aid the sperm's entry into an egg?

12. Where does fertilization usually occur?

13. List, in order, the structures in the female through which the sperm must pass before they reach the egg.

14. In the absence of fertilization, how long does the egg survive?

15. What layers of the secondary oocyte must the sperm penetrate to reach the cell membrane?

16. What marks the beginning of fertilization?

17. What events prevent fertilization of the egg by multiple sperm?

18. What event triggers the completion of meiosis II by the secondary oocyte?

19. What is the secondary oocyte called when it completes meiosis II?

20. When is fertilization considered complete?

21. What structures are lost as the ovum becomes a zygote?

Matching

Select from the following list the correct term for each definition below.

fraternal twins　　**identical twins**　　**twins**　　　　**cleavage**

morphogenesis　　**conjoined twins**　　**differentiation**　　**growth**

_____ 22. cell divisions without cell growth that occur in the first four days after fertilization

_____ 23. an increase in size that begins with implantation

_____ 24. the process of physical changes that occur during development

_____ 25. twins that arise from the ovulation of more than one oocyte

_____ 26. the most common form of multiple births

_____ 27. the development of specialized forms and functions in individual cells

_____ 28. twins that arise when cells begin, but do not complete separation—twins remain attached to each other

_____ 29. twins that arise from a single zygote when cells break apart

21.3   Pre-embryonic development: The first two weeks

21.4   Embryonic development: Weeks three to eight

Labeling

Use the terms below to label each structure in Figure 21.2.

a. pre-embryo

b. trophoblast

c. endoderm

d. diploid nucleus

e. yolk sac

f. ectoderm

g. inner cell mass

h. morula

i. trophoblast cells

j. zygote

k. embryonic disc

l. blastocyst

m. amniotic cavity

n. embryo

o. hollow cavity

p. endoderm

q. trophoblast cells

r. amniotic cavity

s. embryonic disc

t. mesoderm

u. ectoderm

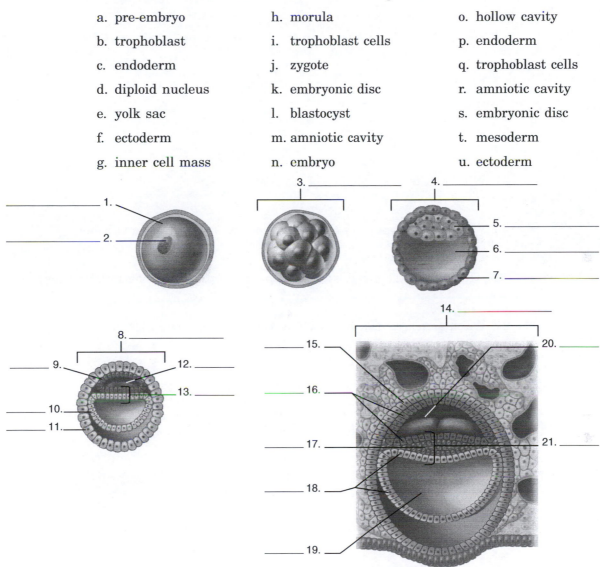

Figure 21.2

Short Answer

Refer to Figure 21.2 as you answer the following questions.

22. What time period after conception is considered to be the pre-embryonic stage?

23. Which structure in Figure 21.2 leaves the oviduct and enters the uterus?

24. Which structure in Figure 21.2 is the first to undergo differentiation and morphogenesis?

25. Which part of the blastocyst will eventually become the embryo?

26. Which structure in Figure 21.2 burrows into the endometrium of the uterus?

27. What is this burrowing process called?

28. What part of the blastocyst secretes enzymes that digest endometrial cells?

29. What marks the end of the pre-embryonic stage?

30. When is the fetus referred to as an embryo?

31. What marks the beginning of the embryonic stage?

32. Which germ layer gives rise to muscle, connective tissue, and bone?

33. Which germ layer gives rise to the lining of the digestive tract and several glands?

34. Which germ layer gives rise to the nervous system and the epidermis of the skin?

Labeling/Coloring

Select a different color for each structure listed below. Color the coding circles and the corresponding structures in Figure 21.3.

○ uterus　　　　　　　○ umbilical cord

○ yolk sac　　　　　　○ amniotic cavity

○ amnion　　　　　　　○ placenta

Figure 21.3

True/False

Mark each statement as true or false. If you mark a statement as false, replace the underlined term or phrase with the term or phrase needed to make the statement true.

_____ 35. Extra-embryonic membranes and the placenta develop early in the <u>embryonic</u> stage.

_____ 36. The four extra-embryonic membranes are the amnion, allantois, <u>yolk sac</u>, and chorion.

_____ 37. The <u>allantois</u> is known as the "bag of waters."

_____ 38. The <u>chorion</u> is a temporary membrane that helps form the blood vessels of the umbilical cord.

_____ 39. The <u>yolk sac</u> provides the germ cells that will move into the gonads and eventually give rise to eggs and sperm.

_____ 40. The placenta forms from the <u>allantois</u> of the fetus and the endometrium of the mother.

21.5   Gender development begins at six weeks

21.6   Fetal development: Nine weeks to birth

21.7   Birth and the early postnatal period

21.8   From birth to adulthood

21.9   Aging takes place over time

21.10   Death is the final transition

Completion

For each event in the development of an embryo/fetus, write in the appropriate week or month during which the event occurs.

_____ 1. the fetus begins moving

_____ 2. the developing embryo is now called a fetus

_____ 3. the fetal face nears its final form

_____ 4. with excellent neonatal care the fetus can survive outside the uterus

_____ 5. the fetal heartbeat can be heard with a stethoscope

_____ 6. male and female internal and external genitalia begin to develop

_____ 7. the testes descend into the scrotal sac in a a male fetus

_____ 8. fetal liver and bone marrow begins producing red blood cells

_____ 9. fetal kidneys begin producing urine

Sequencing

Place the following events leading up to labor in their proper sequence, numbering them from 1 through 6.

_____ 10. fetal adrenal gland releases steroid hormones

_____ 11. fetal pituitary gland matures

_____ 12. the uterus is stimulated to contract rhythmically

_____ 13. the placenta increases production of estrogen and decreases production of progesterone

_____ 14. the number of oxytocin receptors increases

_____ 15. fetal pituitary gland releases ACTH

Fill-in-the-Blank

Referenced sections are in parentheses.

16. A(n) _____ is a surgical incision that enlarges the vaginal opening for birth. (21.6)

17. Pulmonary _____ produced by the infant reduces surface tension in the _____. (21.6)

18. In the fetus, blood flows from the umbilical vein to the inferior vena cava through the _____ _____. (21.6)

19. In the fetal circulation, blood flows from the right atrium to the left atrium through the _____ _____. (21.6)

20. After birth, the umbilical blood vessels regress to become _____ tissue. (21.6)

21. During the first few days after birth, the mother produces a watery milk called _____. (21.6)

22. In the mother, the hormone _____ stimulates milk production and the hormone _____ stimulates delivery of milk. (21.6)

23. The first month of life is referred to as the _____ period. (21.7)

24. During the first 15 months of life, most infants will _____ their weight. (21.7)

25. During infancy, the _____ system is the slowest to develop. (21.7)

26. The brain grows to 95% of its final size during _____. (21.7)

27. In adolescence, puberty is triggered when the hypothalamus begins secreting _____. (21.7)

28. During the period of _____, a lumbar curve develops and abdominal muscles become stronger. (21.7)

29. A section of disposable DNA found on the end of every DNA strand and believed to play a role in aging is called a _____. (21.8)

30. Menopause occurs when the ovaries lose their responsiveness to _____ and _____. (21.8)

31. Describe in one or two words the changes that occur in each of the following with age: (21.8)

   a. lung tissue _____

   b. heart walls _____

   c. bone mass _____

   d. muscle mass _____

   e. lens of the eye _____

   f. viable sperm produced _____

   g. kidney mass _____

   h. prostate gland _____

32. Death is defined as the _____ cessation of circulatory and respiratory, or brain, function. (21.9)

Short Answer

33. What is the role of the SRY gene?

34. What triggers development of female genitalia?

35. What are the three phases of labor and delivery?

Critical Thinking

36. Occasionally, an embryo with a Y chromosome either fails to produce testosterone, or the tissues are insensitive to testosterone. Will this embryo develop as a male or a female? Explain your answer.

Looking Back/Looking Forward

1. Chapter 4 first introduced the negative feedback control system. How does the positive feedback control system of labor and delivery differ from a negative feedback control system?

2. Chapter 23 describes the concept of "growth rate" in a population—the number of births per year minus the number of deaths per year. In the last century, the average lifespan in the United States increased from the 40s to the 70s. What result has this had on the growth rate in our population?

Chapter Test

Multiple Choice

1. In the female reproductive tract, sperm can survive for
   _____ and the egg survives for _____.
   a. one day, one day
   b. several days, one day
   c. one day, several days
   d. several days, several days

2. The protective layer surrounding the secondary oocyte is the
   a. follicle.
   b. granulosa.
   c. corona radiata.
   d. zona pellucida.

3. Fertilization is considered complete when
   a. the sperm enters the egg.
   b. the zona pellucida breaks down.
   c. the nuclei of the sperm and ovum unite.
   d. the sperm attaches to the cell membrane of the secondary oocyte.

4. Ovulation of more than one oocyte in a cycle may result in
   a. fraternal twins.
   b. maternal twins.
   c. conjoined twins.
   d. identical twins.

5. The primary cause of morphogenesis is
   a. cleavage.
   b. differentiation.
   c. fertilization.
   d. growth.

6. Which of the following is true of the blastocyst?
   a. It develops during the embryonic stage.
   b. It is a ball of 32 identical cells.
   c. It consists of a trophoblast and an inner cell mass.
   d. It develops into a morula.

7. Which of the following does not occur during cleavage?
   a. a series of cell divisions
   b. cell differentiation
   c. cell growth
   d. b and c
   e. a, b, and c all occur during cleavage

8. Implantation in the endometrium occurs when
   a. trophoblast cells secrete enzymes that digest endometrial cells.
   b. contact of the blastocyst with the endometrium initiates the growth of new blood vessels.
   c. the morula leaves the oviduct and enters the uterus.
   d. fluid from the amniotic cavity contacts endometrial cells and triggers burrowing of the blastocyst.

9. Muscle and connective tissue arise from which germ layer?
   a. endoderm
   b. ectoderm
   c. mesoderm
   d. trophoderm

10. The placenta is formed during embryonic development from the
    a. chorion and endometrium.
    b. amnion and chorion.
    c. amnion and endometrium.
    d. chorion and myometrium.

11. As an endocrine organ, the placenta secretes
    a. hCG after the corpus luteum degenerates.
    b. progesterone and estrogen to maintain and promote pregnancy when the corpus luteum degenerates.
    c. LH and FSH to stimulate follicle development in the ovaries.
    d. progesterone to maintain the corpus luteum.

12. The neural groove that forms during embryonic development will become
    a. muscle.
    b. bone.
    c. brain and spinal cord.
    d. heart and blood vessels.

13. It is difficult to distinguish a human embryo from any other vertebrate embryo prior to the _____ week of development.
    a. fourth
    b. sixth
    c. eighth
    d. 12th

14. Male and female gonads develop in the
    a. first trimester.
    b. second trimester.
    c. third trimester.
    d. just before birth.

15. A developing human is called a fetus
    a.   after the pre-embryonic stage.
    b.   at the beginning of the second trimester.
    c.   after the eighth week.
    d.   in the third trimester.

16. Contractions of the uterus during labor
    a.   increase the release of oxytocin.
    b.   decrease the release of oxytocin.
    c.   increase the release of prolactin.
    d.   increase the release of ADH.

17. The hormone _____ stimulates milk production in the mother.
    a.   oxytocin
    b.   estrogen
    c.   progesterone
    d.   prolactin

18. Surfactant is important in
    a.   initiating uterine contractions prior to birth.
    b.   preventing blood loss when the umbilical cord is severed.
    c.   redirecting fetal circulatory pathways in the newborn.
    d.   reducing surface tension in the alveoli of the lungs.

19. A shunt that directs fetal circulation from the pulmonary artery to the aorta is the
    a.   ductus arteriosus.
    b.   ductus venosus.
    c.   foramen ovale.
    d.   pulmonary stenosis.

20. Colostrum
    a.   is a pasty substance that covers the newborn's skin.
    b.   is a watery milk produced by the breast.
    c.   is a fluid lost with the afterbirth.
    d.   is an enzyme produced by the myometrium during labor.

21. The period of life from age 2–15 months is termed
    a.   neonatal.
    b.   childhood.
    c.   infancy.
    d.   adolescence.

22. The _____ system lags behind most other systems in its development.
    a. cardiovascular
    b. immune
    c. muscular
    d. nervous

23. All of the following are accepted theories of the cause of aging except
    a. genetic programming determines the timing of cell death.
    b. environmental toxins have led to a shortened life span.
    c. accumulated cell damage interferes with cell repair.
    d. interdependence of body systems promotes aging when one system begins to fail.

24. The finding that severe caloric restriction in some animals prolongs life supports the theory that aging is due to
    a. accumulated damage to DNA and errors in DNA replication.
    b. an internal genetic counting mechanism, possibly involving telomeres, that limits the number of cell divisions in each cell.
    c. the decline in one body system that affects other body systems.
    d. poor nutrition.

25. Death may be defined as a cessation of either circulatory or respiratory function or of brain function. In either definition, a key word preceding "cessation" is
    a. complete.
    b. prolonged.
    c. irreversible.
    d. acute.

Self Test

Each of the Key Concepts listed at the beginning of this chapter has been rewritten as a question below. After successfully completing the study guide exercises and the Chapter Test, you should be able to answer each of these questions. Refer to the Key Concepts list at the beginning of this chapter to check your answers.

1. Trace the path of a sperm from the time it enters the female reproductive tract until it reaches the egg.

2. When does the secondary oocyte complete meiosis II?

3. Where does fertilization usually occur? What is formed by the fusion of the nuclei of the sperm and the ovum?

4. What are the three types of twins?

5. What are the four processes involved in development?

6. What are the three stages of prenatal development?

7. What are the developmental structures that occur in the pre-embryonic stage? Which one becomes implanted in the endometrium?

8. What are the three germ layers that develop during the embryonic stage?

9. Name the four extra-embryonic membranes that develop during the embryonic stage.

10. What structures make up the placenta?

11. What two structures are connected by the umbilical cord?

12. The embryo is distinctly human in appearance by the end of what prenatal stage?

13. What are some of the events that occur during fetal development?

14. What are the three stages of labor and delivery?

15. Is the fetal circulatory system capable of sustaining breathing in the newborn?

16. What substances are provided to the newborn during lactation?

17. What are the five stages of human development that occur after birth?

18. What is aging?

19. Aging is marked by changes in what parts of the body?

20. What is the definition of death?

22

Evolution and the Origins of Life

Chapter Summary and Key Concepts

After reading and studying this chapter, you should know the following:

Sections 22.1, 22.2

1. Evolution is the unpredictable and natural process of descent with modification over time.

2. Microevolution occurs as a result of genetic changes.

3. Macroevolution involves large-scale trends in evolution among groups of species.

4. Charles Darwin proposed the theory of descent with modification in the 1800s.

5. Evidence for evolution comes from the fossil record, comparative anatomy and embryology, biochemistry, and biogeography.

6. Fossils are preserved remnants of organisms.

7. The fossil record is incomplete because fossil formation depends on the type of tissue and the environmental conditions.

8. Homologous structures are similar because they evolved from a common ancestor; analogous structures are similar because they evolved for a similar function.

9. Vestigial structures serve little or no function and may be the remains of an ancestral structure.

10. Similarities in vertebrate embryos suggest that all vertebrates evolved from a common ancestor.

11. Similarities in biochemical molecules provide information about how recently species may have diverged.

12. Biogeography is the study of the distribution of plants and animals around the world.

13. Geographic barriers promote evolution.

14. In the process of natural selection, organisms with traits that make them more fit for the environment are able to survive and reproduce.

15. Evolution is the result of the presence of mutations and the process of natural selection.

16. Genetic drift and gene flow can change populations by changing allele frequencies.

Sections 22.3, 22.4, 22.5

17. Life on Earth began with the formation of organic molecules that became enclosed in lipid-protein membranes.

18. The appearance of photosynthetic organisms allowed oxygen to accumulate in the atmosphere, making possible the eventual appearance of animals.

Section 22.6

19. Humans are classified as *Homo sapiens*.

20. The evolution of *Homo sapiens* required ancestral species to walk upright and develop a larger brain.

Exercises

Complete the exercises for each section after you have read and studied the section. If you cannot answer some questions, or if you answer them incorrectly, return to the chapter and review this information. You may find it helpful to work on only one section at a time. When you have completed all sections, take the Chapter Test as an indicator of your mastery of this topic.

22.1   Evidence for evolution comes from many sources

22.2   Natural selection contributes to evolution

Define

Write the term(s) for each definition below:

_____ 1. the natural process of descent over time with genetic modification

_____ 2. evolution as a result of genetic changes that give rise to new species

_____ 3. large scale evolutionary trends that affect whole groups of species

_____ 4. preserved remnants of organisms

_____ 5. structures found in different species that have similar structure due to descent from a common ancestor

_____ 6. structures found in different species that have a similar function but not a similar structure

_____ 7. structures that may have served a function in an ancestral organism but serve no function in the current organism

_____ 8. the study of the distribution of plants and animals around the world

_____ 9. the slow movement of tectonic plates over time

_____ 10. a group of individuals of the same species that occupy the same geographical area

_____ 11. a general term for random changes in allele frequency due to chance events

_____ 12. a change in allele frequency that occurs when most of a population is lost due to a major catastrophe

_____ 13. a change in allele frequency that occurs when a few individuals leave the original group and become the genetic base of a new population

_____ 14. a change in allele frequency that occurs when individuals move into or out of a population

_____ 15. when a life form dies out completely

_____ 16. a burst of evolutionary activity depicted as numerous branches on an evolutionary tree

Fill-in-the-Blank

Referenced sections are in parentheses.

17. The three components of evolution are _____ _____

_____, _____ _____, and that evolution

is _____ and _____. (Intro)

18. Charles Darwin's phrase for evolution was _____ _____

_____. (Intro)

19. Darwin asserted that life arose _____ time(s). (Intro)

20. Hard tissues are preserved as fossils only when they are quickly covered with

    _____ or _____ _____. (22.1)

21. Sedimentary deposits are laid layer upon layer in a process called

    _____. (22.1)

22. When fossils exist in sedimentary rock, the oldest fossils will be located in the

    _____ layers of rock. (22.1)

23. Radiometric dating in fossils older than 50,000 years requires the use of isotopes of

    _____. (22.1)

24. In humans, the muscles that wiggle the ears are examples of _____

    _____. (22.1)

25. Comparative biochemistry suggests a close relationship between species when they have

    similar or identical _____ _____. (22.1)

26. One species becomes two different species when evolutionary paths

    _____. (22.1)

27. Two hundred million years ago all continents were joined in one land mass called

    _____. (22.1)

Critical Thinking

28. If two different species evolve in the same environment, will this affect the development of homologous or analogous structures? Explain your answer.

29. Common ancestry led to similar embryonic structures in different vertebrate species—although the adult forms are very different. Propose a possible explanation for why the embryonic structures are so similar, whereas the adult structures are not as similar.

22.3   The young Earth was too hot for life

22.4   The first cells were able to live without oxygen

22.5   Photosynthetic organisms altered the course of evolution

Ordering

Label each of the following events with the letters a–k to indicate the sequence in which they occurred as life developed on Earth.

_____   1. Heat vaporized all water, preventing the formation of oceans.

_____   2. Oxygen, a by-product of photosynthesis, begins to accumulate in the atmosphere.

_____   3. Self-replicating RNA molecules form on clay templates along the ocean edge.

_____   4. Multicellular organisms appear.

_____   5. Mutation allows some cells to develop the capacity for photosynthesis.

_____   6. The Earth formed about 4.6 billion years ago with a liquid core of metal and a thin outer crust.

_____   7. Simple organic molecules of amino acids, sugars, and fatty acids form from molecules in the primitive atmosphere, using heat, radiation, and electrical discharges as energy sources.

_____   8. New cells evolve with the capacity to use oxygen for aerobic metabolism.

_____   9. The Earth cools enough for oceans to form.

_____   10. Simple organic compounds dissolve in the sea.

_____   11. RNA and small organic molecules become enclosed within a lipid-protein membrane.

Section 22.6 Modern humans came from Africa

Completion

Fill in the table below by writing the scientific name of each taxonomic category, as it applies to humans. Write one or more characteristics that qualify humans for each category.

Taxonomic Category	Scientific Name	Human Characteristics
Kingdom	1.	2.
Phylum	3.	4.
Class	5.	6.
Order	7.	8.
Family	9.	10.
Genus	11.	12.
Species	13.	14.

Matching

For each of the characteristics listed below, indicate the letter of the species being described.

a. **Homo erectus**   c. **Homo sapiense.**   e. **Ardipithecus ramidus**

b. **Australopithecus**   d. **Homo habilis**
   afarensis

_____ 15. first species believed to have migrated out of Africa

_____ 16. the first distinctly human ancestor

_____ 17. modern humans

_____ 18. the only surviving *Homo* species

_____ 19. the first human ancestor to walk upright

_____ 20. lived about 1.8 million years ago

_____ 21. believed by many scientists to be the first hominid

_____ 22. believed to have been the first toolmakers

_____ 23. the society of this species probably included hunting and gathering groups that shared food

_____ 24. lived about 3.2 million years ago

_____ 25. lived about 2.4 million years ago

_____ 26. crossed the sea to the Arabian peninsula about 50,000 years ago

_____ 27. probably the first to have a diet that included meat

_____ 28. arose about 140,000 years ago

_____ 29. was a contemporary of modern humans

_____ 30. showed the greatest degree of sexual dimorphism

_____ 31. appeared about 4.4 million years ago

Short Answer

32. Two characteristics are not found in Ardipithecus ramidus, causing some scientists to question whether or not they were the first hominids. What are these two characteristics?

33. How do hominoids differ from primates? How do hominids differ from hominoids?

Looking Back/Looking Forward

1. The current issue presented at the beginning of Chapter 22 describes the long, flat feet of the Flores people. Were the genes responsible for these characteristics located on autosomes? (see Chapter 19)

2. Chapter 23 describes what happens when the niches of two different species overlap. Why is this type of competition for resources minimized during adaptive radiation?

Chapter Test

Multiple Choice

1. Evolution that occurs as a result of genetic changes that give rise to a new species is referred to as
   a. microevolution.
   b. macroevolution.
   c. genetic evolution.
   d. allelic evolution.

2. Which of the following is most likely to form a fossil?
   a. a soft tissue organism buried by volcanic ash
   b. an aquatic organism
   c. a vertebrate organism covered by sediment
   d. an insect ingested by a reptile

3. The fossil record is incomplete because
   a. we have no fossil record of many species that lacked hard tissues.
   b. it is difficult to accurately date some fossils.
   c. many people do not accept the evidence for evolution found in the fossil record.
   d. many fossils may be difficult to find.
   e. a and d

4. The forelimbs of all vertebrates are homologous. This means
   a. vertebrates with similar forelimbs evolved in similar environments and experienced the same natural selection processes.
   b. the forelimbs of all vertebrates have identical structures.
   c. all vertebrates descended from a common ancestor.
   d. vertebrate evolution occurred more quickly than that of other organisms.

5. Radioactive dating of fossils younger than 50,000 years uses
   _____ isotopes, while dating of older fossils uses
   _____ isotopes.
   a. oxygen, carbon
   b. carbon, potassium
   c. potassium, magnesium
   d. potassium, carbon

6. The coccyx, or tailbone, of a human is a(n) _____ structure.
   a. homologous
   b. vestigial
   c. analogous
   d. divergent

7. The presence of identical or nearly identical biochemical molecules in two
   species indicates
   a. recent divergence.
   b. the presence of analogous structures.
   c. that they developed in the same geographical region.
   d. an increased rate of mutation.

8. The bottleneck effect results in the survival of individuals who
   a. carry genes representing the most fit genes in the population.
   b. are the best adapted for survival and reproduction.
   c. share a very small gene pool that was selected with no regard to
      fitness.
   d. left the original population and began a new population in a different
      location.
   e. a and b

9. The effect of continental drift has been to
   a. increase the rate of mutation.
   b. decrease the rate of mutation.
   c. isolate related groups of organisms from each other.
   d. maintain Pangaea.

10. A population is a group of individuals
    a. who share the same alleles.
    b. who occupy the same area.
    c. of the same species who occupy the same area.
    d. of the same species, anywhere on Earth.

11. Random changes in the allele frequency because of chance effects is
    called
    a. gene flow.
    b. natural selection.
    c. genetic drift.
    d. genetic change.

12. The largest mass extinction was the
    a. Triassic.
    b. Jurassic.
    c. Cretaceous.
    d. Diluvian.

13. Short bursts of evolutionary activity are called
    a. divergence.
    b. adaptive radiation.
    c. natural selection.
    d. selective radiation.

14. The Earth formed about
    a. 6 billion years ago.
    b. 4.6 billion years ago.
    c. 4 million years ago.
    d. 2 million years ago.

15. Which of the following was not found in the primitive atmosphere of the earth?
    a. CO_2
    b. H_2O
    c. O_2
    d. N_2

16. Before life evolved, simple organic molecules formed from
    a. the oceans.
    b. atmospheric gases.
    c. the rain.
    d. DNA and RNA.

17. The first self-replicating molecule was probably
    a. DNA.
    b. RNA.
    c. a lipid.
    d. a protein.

18. The evolution of photosynthetic organisms resulted in
    a. an increased variety of proteins.
    b. increased oxygen in the atmosphere.
    c. evolution of DNA.
    d. increased carbon dioxide in the atmosphere.

19. Organisms capable of aerobic metabolism evolved
    a. prior to photosynthetic organisms.
    b. after photosynthetic organisms.
    c. at the same time as photosynthetic organisms.
    d. as soon as Earth cooled.

20. Humans and chimpanzees diverged from a common ancestor about
    a. 2 billion years ago.
    b. 5 million years ago.
    c. 50,000 years ago.
    d. 10,000 years ago.

21. In comparing hominids and primates,
    a. both hominids and primates have a tail.
    b. hominids have smaller brains than primates.
    c. both are arboreal.
    d. hominids have more complex social behavior than primates.

22. All of the following apply to *Australopithecus afarensis* except
    a. their diet consisted of vegetables.
    b. they are classified as hominid.
    c. they displayed sexual dimorphism.
    d. they gave rise to *Homo erectus*.

23. Humans qualify for the phylum Chordata because they have
    a. upright posture.
    b. spoken language.
    c. a nerve cord.
    d. mammary glands.

24. Evidence for chimpanzees as our closest relatives comes from
    a. analogous structures.
    b. the absence of a tail in chimpanzees.
    c. similar thumb structure.
    d. similarities in human and chimpanzee DNA.

25. Hominids are characterized by
    a. complex language skills.
    b. opposable thumbs.
    c. upright posture.
    d. mammary glands.

Self Test

Each of the Key Concepts listed at the beginning of this chapter has been rewritten as a question below. After successfully completing the study guide exercises and the Chapter Test, you should be able to answer each of these questions. Refer to the Key Concepts list at the beginning of this chapter to check your answers.

1. What is evolution?

2. What causes microevolution to occur?

3. What characterizes macroevolution?

4. Who proposed the theory of descent with modification?

5. What are four areas that provide evidence for evolution?

6. What are fossils?

7. Why is the fossil record incomplete?

8. What are homologous structures? What are analogous structures?

9. What are vestigial structures?

10. What characteristic of vertebrate embryos supports the theory of evolution?

11. What information is gained by analyzing biochemical molecules?

12. What is biogeography?

13. What effect do geographic barriers have on evolution?

14. What is the process of natural selection?

15. What two things contribute most to the process of evolution?

16. How can genetic drift and gene flow change populations?

17. How did life on Earth probably first begin?

18. What effect did photosynthetic organisms have on the atmosphere of the primitive Earth?

19. What is the genus and species for humans?

20. What two characteristics were required in ancestral organisms for the development of humans?

23

Ecosystems and Populations

Chapter Summary and Key Concepts

After reading and studying this chapter, you should know the following:

Sections 23.1, 23.2, 23.3

1. Ecology is the study of the relationships between organisms and their physical environment.

2. Organisms and environments interact as populations, communities, ecosystems, and the biosphere.

3. Characteristics of populations include habitat, range, size, actual growth rate, and potential for growth.

4. The biotic potential of a population is the maximum rate of growth, and is limited by environmental resistance.

5. The population that an ecosystem can support indefinitely is called the carrying capacity.

6. Rapid growth in the human population began in the 1700s when new developments increased the carrying capacity.

7. The fertility rate is the number of children born to each woman; the replacement fertility rate is the average fertility rate required to reach zero population growth.

8. The fertility rate is not the same in all nations; the least developed countries with the fewest resources have the highest fertility rates.

9. An organism's niche is its role in the community. Competition may result when niches for different species overlap.

10. Succession is the natural sequence of change in a community, and is determined by population growth rates, niches occupied by various species, and competition.

11. Succession ends with the establishment of a climax community.

12. An ecosystem consists of living components called the biomass, and nonliving components made up of chemical elements and energy.

Sections 23.4, 23.5

13. Energy flow through an ecosystem is governed by the laws of thermodynamics.

14. The first law of thermodynamics states that energy cannot be created or destroyed.

15. The second law of thermodynamics states that whenever energy changes form or is transferred, some of the energy will be transformed into nonuseful forms.

16. Producers are organisms capable of photosynthesis.

17. Consumers are organisms that must consume food for energy.

18. In an ecosystem, energy flows from the sun to producers, and then to consumers.

19. In a biogeochemical cycle, chemicals cycle between the biomass, exchange pool, and reservoir pool.

20. Important biogeochemical cycles include the cycles of water, carbon, nitrogen, and phosphorus.

Exercises

Complete the exercises for each section after you have read and studied the section. If you cannot answer some questions, or if you answer them incorrectly, return to the chapter and review this information. You may find it helpful to work on only one section at a time. When you have completed all sections, take the Chapter Test as an indicator of your mastery of this topic.

23.1   Ecosystems: Communities interact with their environment

23.2   Populations: The dynamics of one species in an ecosystem

23.3   Human population growth

Define

Write the term(s) for each definition below:

_____   1. the study of the relationships between organisms and their physical environment

_____   2. a community of organisms and the physical environment in which they live

_____   3. a group of individuals of the same species occupying the same geographical area

_____   4. the populations of all the species occupying a geographical area

_____   5. all the ecosystems of the earth

_____   6. the area in which a species lives

_____   7. the maximum rate of growth of a population under ideal conditions

_____ 8. a limiting force that prevents a species from reaching its biotic potential

_____ 9. the population size that an ecosystem can support indefinitely

_____ 10. the number of births per year minus the number of deaths per year

_____ 11. the point at which birthequals deaths in a population

_____ 12. the study of human populations

Fill-in-the-Blank

Referenced sections are in parentheses.

13. A species' _____ is determined by its tolerance for environmental conditions. (23.2)

14. Three things that limit an organism's range are _____
_____ _____, _____
_____, and _____ _____. (23.2)

15. A population stabilizes when it reaches a balance between _____
_____ and _____ _____. (23.2)

16. Factors that kill organisms or prevent them from reproducing contribute to
_____ _____. (23.2)

17. Every growing population will eventually reach a point where environmental resistance begins to _____. (23.2)

18. The global human population first began to rise slowly with agricultural development, and conditions led to a(n) _____ in environmental resistance and a(n) _____ in carrying capacity. (23.3)

19. In order to reach zero population growth, the human population will have to either _____ the birth rate or _____ the death rate. (23.3)

20. In 2000, the human fertility rate was _____. (23.3)

21. The fertility rate is lowest in _____ industrialized countries. (23.3)

22. In the first stage of demographic transition, a society will have a(n) _____ death rate, and a(n) _____ birth rate. (23.3)

23. In the final stages of demographic transition, the birth rate _____ the death rate. (23.3)

True/False

Mark each statement as true or false. If you mark a statement false, replace the underlined term or phrase with the term or phrase needed to make the statement true.

_____ 24. Characteristics of a <u>species</u> include habitat, range, size, actual growth rate, and potential capacity for growth.

_____ 25. An exponential growth curve is characterized by a <u>J</u> shape.

_____ 26. The human fertility rate is the average number of children born to <u>women who have children</u> in their lifetime.

_____ 27. A demographic transition is a progression of population changes linked to the <u>population growth</u> of a region.

_____ 28. A nation with a well developed industry-based economy is classified by demographers as a <u>LIC</u>.

_____ 29. A species' habitat is determined by the <u>chemical and physical characteristics</u> that maximize its chances for survival.

_____ 30. Carrying capacity <u>can</u> change if environmental conditions change.

Short Answer

31. What types of interactions will occur between animals and plants that occupy the same or overlapping habitats?

32. What four factors influence the biotic potential of a species?

33. What contributed to the stability of the human population prior to 4,000 years ago?

34. What historical event coincides with the beginning of rapid growth in the human population?

35. Why is the replacement fertility rate 2.1 rather than 2.0?

36. Which type of countries has the highest fertility rate? What challenges may result from this?

23.4   Communities: Different species living together

23.5   Energy flows through ecosystems

23.6   Chemical cycles recycle molecules in ecosystems

Matching

Select from the following list the correct term for each definition below.

climax   succession   niche   ecological
  community   pyramid
                biomass   biogeochemical
producers   cycle   photosynthesis
                nitrogen
consumers   fixation   herbivores   carnivores

denitrification omnivores   decomposers   fossil fuels

food web   nitrification

_____ 1. a stable community achieved by succession

_____ 2. indicates total amount of energy at each level of an ecosystem
                or total biomass stored at each level

_____ 3. a step in the biogeochemical cycle of nitrogen in which bacteria
                convert nitrates into atmospheric nitrogen gas

_____ 4. the living components of an ecosystem

_____ 5. the process of converting atmospheric nitrogen gas to ammonia

_____ 6. carbon-containing compounds obtained from fossils and used
                as an energy source

_____ 7. a natural sequence of change in the organisms that dominate
                a community

_____ 8. organisms that cannot synthesize their own food

_____ 9. the interaction of multiple food chains

_____ 10. consumers that utilize the energy stored in green plants

_____ 11. an organism's role in the community

_____ 12. consumers that obtain energy from dead organisms

_____ 13. consumers that can derive energy from plants or animals

_____ 14. the recycling of chemicals that compose living organisms
                between the organisms and the earth

_____ 15. the formation of nitrate

_____ 16. organisms capable of photosynthesis

_____ 17. consumers that feed on other animals

_____ 18. harvesting solar energy to produce organic molecules

Short Answer

19. Climax communities are very efficient in their utilization of what types of
    resources?

20. Why are climax communities slow to recover if they are disrupted?

Labeling

On the left side of Figure 23.1, label the role each organism plays in the food chain. On the right side of the figure, label the characteristics applicable to each organism. Use the lists provided; each characteristic may be used more than once.

Roles

a.   secondary consumer

b.   producer

c.   tertiary consumer

d.   primary consumer

Characteristics

a.   referred to as an autotroph

b.   release CO_2 into the atmosphere

c.   herbivore

d.   performs photosynthesis

e.   release O_2 into the atmosphere

f.   carnivore

g.   referred to as a heterotroph

h.   least efficient use of energy

_____ 21.   25. _____

_____ 22.   26. _____

_____ 23.   27. _____

_____ 24.   28. _____

Figure 23.1

Use the phrases below to label Figure 23.2 of the water cycle.

a.   precipitation over land   d.   evaporation from land

b.   precipitation over ocean   e.   evaporation from ocean

c.   ocean to land   f.   land to ocean

Figure 23.2

Fill-in-the-Blank

35. The first law of thermodynamics states that energy cannot be created or

_____. (23.5)

36. Small organelles called _____ are found in plant cells and house the

reactions of photosynthesis. (23.5)

37. Plants contain a light-absorbing pigment called _____. (23.5)

38. In the first stage of photosynthesis, sunlight is absorbed, _____ and

_____ are produced, and _____ is released. (23.5)

39. The_____ cycle occurs in the second stage of photosynthesis. (23.5)

40. The main product of photosynthesis is high energy sugar molecules built with

_____ from the sun, and carbon atoms from _____

_____. (23.5)

41. During the Calvin cycle of photosynthesis, plants use CO_2 from the atmosphere to manufacture _____. (23.5)

42. In a biogeochemical cycle, producers obtain nutrients from the _____ pool, and incorporate them into the _____. (23.6)

43. The biogeochemical cycle for carbon is called a _____ cycle due to the exchange of carbon between living organisms and the atmosphere. (23.6)

44. Nitrogen fixation is carried out by _____. (23.6)

45. All organisms, other than plants, must rely on _____ for usable nitrogen. (23.6)

46. In denitrification, denitrifying bacteria convert some nitrates to _____ _____. (23.6)

47. The phosphorus cycle is called a _____ cycle because phosphorus never enters the atmosphere. (23.6)

Critical Thinking

48. Why is energy described as flowing through the ecosystem in one direction, rather than being recycled?

49. Why are nutrients described as cycling through the ecosystem, rather than flowing in one direction?

Looking Back/Looking Forward

1. The current chapter mentions that human activity may inadvertently lower the carrying capacity of the earth. Chapter 20 described how transgenic plants may increase the carrying capacity of many regions, and decrease the death rate due to disease. Do you think that one of these factors will have greater effect than the other on our human population? Explain your answer.

2. Chapter 24 describes many ways in which human activity is changing our ecosystems, depleting resources, and decreasing biodiversity. Will this increase or decrease environmental resistance?

Chapter Test

Multiple Choice

1. Ecology is best defined as the study of
   a. the cycling of elements in the environment.
   b. the relationship between organisms and their physical environment.
   c. ecological pyramids.
   d. fossil fuels.

2. Which of the following does not limit an organism's geographic range?
   a. competition for resources
   b. reproductive behaviors
   c. extreme or intolerable conditions
   d. physical obstacles and geographic barriers

3. The biotic potential of an organism will usually
   a. increase without limits.
   b. increase independently of carrying capacity.
   c. not be affected by environmental resistance.
   d. follow an exponential growth curve.

4. One determinant of the biotic potential of a population is
   a. time required for offspring to reach reproductive maturity.
   b. genetic compatibility.
   c. gene flow and genetic drift.
   d. environmental conditions and geographic barriers.

5. The rapid rise in human population can be attributed to an increase in all of the following except
   a. industrialization.
   b. availability of antibiotics and vaccines.
   c. environmental resistance.
   d. geographic range.

6. Industrialization raises the _____ of a human population.
   a. biotic potential
   b. environmental resistance
   c. reproductive rate
   d. carrying capacity

7. Environmental resistance is best defined as
   a. the ability of an organism to adapt to a wide range of environments.
   b. the ability of an organism to overcome environmental obstacles to reproduction.
   c. factors that limit a species' ability to realize its biotic potential.
   d. the number of populations competing for the same resources.

8. The growth rate of a population is calculated as the number of births per year *minus* the number of deaths per year *divided* by the
   a.   fertility rate.
   b.   replacement fertility rate.
   c.   current growth rate.
   d.   population size.

9. A population can fail to have zero population growth even after reaching the replacement fertility rate due to the
   a.   age structure.
   b.   infant mortality rate.
   c.   number of elderly in the population.
   d.   environmental conditions.

10. Where do we see the highest growth rates in human populations?
    a.   MICs
    b.   LICs
    c.   there is no distinguishable pattern
    d.   countries with higher economic and educational levels

11. The population that an environment can support indefinitely is called the
    a.   climax community.
    b.   carrying capacity.
    c.   maximum population.
    d.   environmental potential.

12. The carrying capacity of an ecosystem is achieved when
    a.   a balance occurs between biotic potential and environmental resistance.
    b.   the reproductive rate of a population declines.
    c.   environmental resistance increases.
    d.   competitive exclusion increases.

13. Competitive exclusion could not occur without
    a.   two or more populations competing for the same resources.
    b.   two or more organisms of the same species competing for the same resources.
    c.   more organisms of reproductive age than the environment is capable of supporting.
    d.   a change in the physical environment that limits the availability of resources.

14. In photosynthesis, the products of the Calvin cycle include
    a. O_2 and CO_2.
    b. O_2 and sugar.
    c.   sugar and ATP.
    d.   sugar and ADP.

15. Chlorophyll is
    a. a plant cell that carries out photosynthesis.
    b. a species of algae that functions as a producer in an aquatic ecosystem.
    c. a light-absorbing pigment.
    d. a storage organelle for excess carbohydrates in plant cells.

16. Climax communities are
    a. the most efficient in energy and nutrient utilization.
    b. occupied by one species.
    c. not involved in competition.
    d. only found in certain locations.

17. A constant supply of energy is essential for continued life because the transfer of energy from one organism to another organism is an energy-_____ process.
    a. absorbing
    b. producing
    c. releasing
    d. balanced

18. The chemical matter of the Earth is _____ the Earth and the biomass.
    a. transferred to
    b. divided between
    c. produced by
    d. recycled between

19. The second law of thermodynamics states that
    a. energy cannot be created or destroyed.
    b. energy can be changed from one form to another.
    c. energy can be used, or stored in a usable form.
    d. energy transfer is inefficient.

20. Humans can function as all of the following except
    a. secondary consumers.
    b. tertiary consumers.
    c. omnivores.
    d. autotrophs.

21. The population of consumers at any level of a food chain depends most directly on
    a. the amount of energy available from the sun.
    b. the biogeochemical cycles of essential chemicals.
    c. the population of consumers directly below it.
    d. the vulnerability of the producers.

22. Molecules and elements in living things cycle between the biomass, a reservoir, and a(n)
    a. producer.
    b. consumer.
    c. exchange pool.
    d. recycling pool.

23. Which of the following would not affect the carbon cycle?
    a. photosynthesis by plants
    b. aerobic respiration by plants and animals
    c. sunlight
    d. atmospheric phosphorus

24. Living things depend on nitrogen fixation
    a. to detoxify atmospheric nitrogen.
    b. to convert atmospheric nitrogen to a usable form.
    c. to make nitrogen available for aerobic respiration.
    d. to return nitrogen to the atmosphere.

25. The phosphorus cycle is a sedimentary cycle because
    a. phosphorus cycles through only two pools.
    b. phosphorus is obtained from sediments.
    c. the phosphorus cycle relies on water availability.
    d. phosphorus never enters the atmosphere.

Self Test

Each of the Key Concepts listed at the beginning of the study guide has been rewritten as a question below. After successfully completing the study guide exercises and the Chapter Test, you should be able to answer each of these questions. Refer to the Key Concept List at the beginning of this chapter to check your answers.

1. What is ecology?

2. List four levels of interaction involving organisms with each other or with the environment.

3. List five characteristics of populations.

4. What is the biotic potential of a population and what limits this potential?

5. What is carrying capacity?

6. When did rapid growth in the human population begin, and what made this growth possible?

7. Define fertility rate and replacement fertility rate.

8. In what types of countries are fertility rates the highest?

9. What is an organism's niche, and what happens when the niches for two different species overlap?

10. What is succession, and what factors influence it?

11. What is the culmination of succession?

12. What are the living and nonliving components of an ecosystem?

13. What laws govern the flow of energy through an ecosystem?

14. What is the first law of thermodynamics?

15. What is the second law of thermodynamics?

16. What are producers?

17. What are consumers?

18. What is the sequence of energy flow in an ecosystem between producers, consumers, and the sun?

19. What happens in a biogeochemical cycle?

20. List four common and important chemicals involved in biogeochemical cycles.

Human Impacts, Biodiversity, and Environmental Issues

Chapter Summary and Key Concepts

After reading and studying this chapter, you should know the following:

Sections 24.1, 24.2, 24.3

1. Pollution affects our air, water, and land resources.

2. The effects of air pollution include global warming, destruction of the ozone layer, acid precipitation, and smog production.

3. The primary greenhouse gases are carbon dioxide, methane, and the air pollutants CFCs.

4. The greenhouse effect allows sunlight to pass through the atmosphere while trapping heat below. This is a normal process and helps to maintain the temperature of the earth.

5. Global warming is an increase in average global temperatures and occurs when excess greenhouse gases are produced.

6. Two human activities that contribute to the potential for global warming are deforestation and the burning of fossil fuels.

7. Ozone occurs near the earth's surface as a result of air pollution and in the stratosphere where it is protective. CFCs deplete the stratospheric ozone layer.

8. Acid precipitation is produced when air pollutants combine with water in the atmosphere.

9. Smog occurs when air pollutants react with each other and usually appears after an area has experienced rapid industrial growth.

10. The negative effects on water of human activities include excessive use, promoting runoff, and increasing water pollution.

11. Toxic pollutants in water impact biological magnification, concentrating toxic substances in tertiary consumers.

12. Human activities can lead to pollution of surface water, ground water, and ocean water.

13. Human activities contribute to about 50% of oceanic oil pollution.

14. The expansion of human populations and the search for resources has polluted existing land and led to a reduction in productive land.

15. Challenges that arise when cities replace farmlands include an increased demand on resources and the loss of farming productivity.

16. Land damage in rural areas occurs through stripping of resources by local populations.

Sections 24.4, 24.5, 24.6

17. Because energy use is determined by consumption, we have the power to make choices as to where we obtain energy and how we use it.

18. Humans have high energy demands and inadequate energy resources.

19. Current energy sources are primarily fossil fuels; additional energy sources include nuclear energy, hydroelectric power, biomass fuel, and solar energy.

20. A sustainable world is one in which both humans and the ecosystems are maintained.

Exercises

Complete the exercises for each section after you have read and studied the section. If you cannot answer some questions, or if you answer them incorrectly, return to the chapter and review this information. You may find it helpful to work on only one section at a time. When you have completed all sections, take the Chapter Test as an indicator of your mastery of this topic.

24.1   Pollutants impair air quality

24.2   Pollution jeopardizes scarce water supplies

24.3   Pollution and overuse damage the land

Define

Write the term(s) for each definition below.:

_____   1. chemicals in the environment that have adverse effects on living organisms

_____   2. when sunlight is allowed to pass through the atmosphere while most of the heat is trapped below

_____   3. a rise in the average global temperature

_____   4. the removal of trees from large areas of land

_____   5. O_3

_____   6. the transformation of productive land into desert

_____   7. sulfuric acid and nitric acid dissolved in raindrops

_____   8. a brown or gray layer formed in the atmosphere when air pollutants react with each other in the presence of sunlight and
water vapor

_____   9. a condition that occurs when a warm upper layer of air traps a
cooler air mass containing smog beneath it

_____ 10. deep underground reservoirs of fresh water

_____ 11. when the accumulation of organic and inorganic nutrients in a
shallow body of water leads to the growth of plant life and the
death of animal life

_____ 12. a condition in which toxic pollutants become increasingly concentrated in the tissues of animals higher in up the food chain

Short Answer

Refer to Figure 24.1 to answer the following questions.

Figure 24.1

13. Which pathways of the carbon cycle have been affected by human activity in such a way that atmospheric levels of CO_2 are increasing?

14. Describe the activity associated with each pathway and how it is causing CO_2 levels to rise.

15. Does this increase in CO_2 levels increase or decrease the greenhouse effect?

16. What effect could rising CO_2 levels have on the temperature of the earth?

Word Choice

Circle the answer that correctly completes the sentence.

17. Ozone in the (troposphere/stratosphere) helps to protect the earth from UV radiation.

18. Ozone in the (troposphere/stratosphere) is a pollutant formed when oxygen reacts with automobile exhaust and industrial pollution.

19. CFCs (increase/decrease) the level of ozone in the stratosphere.

20. CFCs (increase/decrease) global warming.

21. Acid precipitation may form when (chlorine/sulfur dioxide) is released into the air.

22. Nitrogen oxides and hydrocarbons are the primary contributors to (global warming/smog formation).

23. Freshwater makes up (less than/more than) 1% of the Earth's total amount of water.

24. Water pollutants that are organic nutrients (increase/decrease) the growth rate of bacteria, resulting in a(n) (increased/decreased) availability of oxygen for other aquatic organisms.

25. Eutrophication (converts freshwater into/protects freshwater from becoming) a marsh.

26. Most toxic pollutants (can/cannot) be degraded by biological decomposition.

27. In biological magnification, toxic substances become (more/less) concentrated in the tissues of animals higher up in the food chain.

28. Many of the pollutants that affect surface water (do/do not) pollute groundwater.

Fill-in-the-Blank

Referenced sections are in parentheses.

29. Air pollution impacts the environment in four areas: _____

    _____, _____ _____,

    _____ _____, and _____

    _____. (24.1)

30. CFCs produce _____ atoms that convert ozone into _____.

    (24.1)

31. Sewage treatment plants, food-packing plants, and paper mills produce

    _____ _____ which pollutes water. (24.2)

32. The pollution of water by inorganic nutrients causes an increase in the growth of

    _____. (24.2)

33. Heat pollution affects water by decreasing the amount of _____ that water

    can carry. (24.2)

34. One source of heat pollution is _____ _____. (24.2)

35. Pollution of groundwater is a concern for human health because groundwater is often a source

    of _____ _____. (24.2)

36. The EPA estimates that up to _____ percent of all water systems and rural

    wells are polluted. (24.2)

37. About _____ percent of oceanic oil pollution is caused by people. (24.2)

24.4 Energy: Many options, many choices

24.5 Human impacts are creating a biodiversity crisis

24.6 Sustainable development supports future economic needs

Completion

Fill in the table below with the disadvantages of utilizing the various sources of energy listed.

Energy Source	Material Required	Disadvantages
Fossil Fuels	1.	2.
		3.
		4.
Nuclear Energy	Uranium	5.
		6.
Hydroelectric Power	7.	8.
Biomass Fuels	9.	10.
		11.
		12.
		13.

Short Answer

14. Solar power is a renewable energy source. What currently prevents solar power from becoming our primary energy source? (24.4)

15. What is one measurement of biodiversity? Which human activities have contributed to a reduction in biodiversity? (24.5)

16. How did the 1987 World Commission on Environment and Development define "sustainable development"? (24.6)

17. Describe the differences between the GDP and GPI. (24.6)

18. List several strategies that can be used in supporting sustainable development. (24.6)

Looking Back

1. Fig. 23.10 depicts the energy stored in the biomass at each producer/consumer level, with tertiary consumers storing the least amount of energy. Chapter 24 describes the process of biological magnification, in which tertiary consumers accumulate the greatest amount of toxic pollutants. How can tertiary consumers accumulate the highest levels of pollutants and the lowest amount of energy?

Chapter Test

Multiple Choice

1. Which of the following is not a primary component of the air we breathe?
   a. nitrogen
   b. oxygen
   c. carbon dioxide
   d. magnesium

2. The greenhouse effect
   a. prevents sunlight from passing through the atmosphere.
   b. allows heat to escape through the atmosphere.
   c. is important in maintaining normal surface temperatures on the earth.
   d. is not involved in global warming.

3. The two main human activities that raise the level of CO_2 in the atmosphere are
   a. deforestation.
   b. desertification.
   c. burning of fossil fuels.
   d. a and c.
   e. b and c.

4. Ozone in the atmosphere is
   a. beneficial in the troposphere and harmful in the stratosphere.
   b. beneficial in the stratosphere and harmful in the troposphere.
   c. beneficial in both the troposphere and the stratosphere.
   d. harmful in both the troposphere and the stratosphere.

5. CFCs contribute to global warming by
   a. releasing chlorine atoms that produce more ozone.
   b. destroying the chlorine atoms that normally produce ozone.
   c. releasing chlorine atoms that destroy ozone and produce oxygen.
   d. releasing oxygen atoms that convert ozone into chlorine atoms.

6. Which of the following does not contribute directly to the formation of acid precipitation?
   a. burning high-sulfur coal
   b. automobile exhaust
   c. burning oil
   d. deforestation

7. Acid precipitation is produced when
   a. holes in the ozone layer affect atmospheric gases.
   b. sulfur dioxide and nitrogen oxides combine with water vapor in the air.
   c. water in the atmosphere ionizes.
   d. industrial plants release acid vapors into the air.

8. The two factors that contribute most to smog are
   a. acid rain and the burning of fossil fuels.
   b. the burning of fossil fuels and automobile exhaust.
   c. an increase in greenhouse gases and automobile exhaust.
   d. eutrophication and automobile exhaust.

9. What condition can be worsened by a thermal inversion?
   a. ozone depletion
   b. smog
   c. acid rain
   d. global warming

10. Freshwater makes up _____ of the Earth's total water.
    a. less than 1%
    b. less than 5%
    c. more than 10%
    d. more than 50%

11. The availability of freshwater in the world
    a. is used primarily by populations of more industrialized countries.
    b. is used primarily by populations of less industrialized countries.
    c. is equally distributed throughout most nations.
    d. has little or no effect on human carrying capacity.

12. Water pollutants that are inorganic nutrients
    a. cause prolific growth of algae.
    b. cause prolific growth of bacteria.
    c. are degraded by bacteria.
    d. arise from sewage treatment plants.

13. Eutrophication
    a. refers to the rapid growth of animal life in a shallow body of water.
    b. results from a depletion of organic nutrients.
    c. is accelerated by water pollution.
    d. converts dry land into marsh.

14. In biological magnification, _____ become more concentrated in animals that are _____ .
    a. toxic substances, higher up in the food chain
    b. toxic substances, lower down in the food chain
    c. nutrients, higher up in the food chain
    d. nutrients, lower down in the food chain

15. Which of the following does not contribute directly to water pollution?
    a. disease-causing organisms
    b. heat
    c. sediment
    d. CFCs

16. Heat pollution in water
    a. increases the amount of oxygen in the water.
    b. decreases the oxygen demand of some aquatic organisms.
    c. may lead to suffocation of some aquatic organisms.
    d. protects freshwater from organic contaminants.

17. It is difficult to determine the full effects of groundwater contamination because
    a. the effects may be confused with other causes of human disease.
    b. it is difficult to detect the presence of radioactive waste.
    c. groundwater is protected from most contaminants by its location.
    d. many effects of groundwater contamination are never reported.

18. About half of the oil that enters the ocean every year comes from
    a. oil disposal on land that washes out to sea.
    b. accidents at sea.
    c. natural seepage.
    d. acid rain.

19. Which of the following contributes least to the damage and overuse of land?
    a. reliance on the land as a resource in rural communities
    b. expansion of cities
    c. urban poverty
    d. war

20. When land becomes unusable for agricultural purposes as a result of approaching desert-like conditions, this is called
    a. desertification.
    b. erosion.
    c. heat pollution.
    d. eutrophication.

21. Which of the following is a true statement about fossil fuels?
    a. They are one fuel source that does not have an environmental cost.
    b. Use of fossil fuels may reduce the level of greenhouse gases.
    c. Use of fossil fuels may contribute to acid rain.
    d. They are a nonrenewable energy source.

22. Examples of biomass fuels include
    a. wood.
    b. coal.
    c. oil.
    d. water.

23. The original renewable energy source for our world is
    a. hydroelectric power.
    b. solar energy.
    c. human kinetic energy.
    d. biomass fuels.

24. Human activities that impact biodiversity include all of the following except
    a. heat pollution.
    b. depletion of natural resources.
    c. farming.
    d. deforestation.
    e. All forms of pollution and resource depletion impact biodiversity

25. Which of the following would not contribute to the establishment of a sustainable world?
    a. recycling
    b. reducing consumption
    c. increasing the fertility rate
    d. reducing poverty

Self Test

Each of the Key Concepts listed at the beginning of the study guide have been rewritten as a question below. After successfully completing the study guide exercises and the Chapter Test, you should be able to answer each of these questions. Refer to the Key Concept List at the beginning of this chapter to check your answers.

1. Which areas of our environment are affected by pollution?

2. What are the four primary effects of air pollution?

3. What are the primary greenhouse gases?

4. What is the greenhouse effect? Is it a normal process? What aspect of the environment do greenhouse gases help to maintain?

5. What is global warming? What may cause global warming?

6. Which human activities contribute to the potential for global warming?

7. Where is ozone found? In which area is ozone beneficial? What chemicals contribute to depletion of the ozone layer?

8. What causes acid precipitation?

9. What causes smog? When does it usually appear?

10. List three ways in which human activities affect water as a resource.

11. What happens in the process of biological magnification?

12. What sources of water can by polluted by human activities?

13. What percentage of oceanic oil pollution is the result of human activity?

14. What effect have human activities had on existing land and available productive land?

15. List two challenges that arise when cities replace farmlands.

16. What is the primary cause of land damage in rural areas?

17. In what ways do we have the power to make choices in where we obtain energy and how we use it?

18. How would you characterize the energy demands of human populations?

19. What is currently the primary energy source for human populations? List four additional energy sources.

20. What is a sustainable world?

Answer Key

Chapter 1

Sections 1.1, 1.2

1. science; **2.** biology; **3.** metabolism; **4.** homeostasis; **5.** multicellular; **6.** DNA; **7.** Monera; **8.** Animalia; **9.** Plantae; **10.** Fungi; **11.** Protista; **12.** Eukarya; **13.** Bacteria, Archaea; **14.** species; **15.** genus; **16.** carbohydrates, lipids, proteins, nucleic acids; **17.** photosynthesis; **18.** matter, energy; **19.** eukaryotic, prokaryotic; **20.** reproduce; **21.** vertebrates; **22.** mammals; **23.** primates; **24.** bipedalism; **25.** opposable; **26.** brain; **27.** language; **28.** atom; **29.** molecule; **30.** cell; **31.** tissue; **32.** organ; **33.** organ system; **34.** organism; **35.** population; **36.** community; **37.** ecosystem; **38.** biosphere; **39.** Refer to section 1.1.; **40.** While individual organisms may not show every characteristic, populations of similar organisms do.; **41.** Domain encompasses kingdoms.; **42.** The smallest unit of life, exhibiting the characteristics of life, is a cell; since an organelle is only a part of a cell it does not qualify as living.; **43.** Every day human beings breathe and bring water into their bodies, obtaining raw materials such as oxygen and hydrogen atoms from these nonliving substances. Energy is obtained from the foods we eat, such as plant and animal material, which are living substances.; **44.** Homeostasis and responding to the external environment are similar because both processes seek to provide conditions favorable to life. They are different because homeostasis manages the internal environment inside your body, while responding to the external environment helps you interact in a healthy way with the world outside your body.

Sections 1.3, 1.4, 1.5, 1.6

1. theory; **2.** placebo; **3.** variable; **4.** technology; **5.** control group; **6.** skepticism; **7.** scientific knowledge; **8.** hypothesis; **9.** experiment; **10.** experimental group; **11.** scientific method; **12.** data; **13.** F, process; **14.** F, inductive; **15.** T; **16.** F, prediction; **17.** T; **18.** F, blind; **19.** F, hypothesis; **20.** T; **21.** F, peer-reviewed; **22.** F, abscissa; **23.** T; **24.** T; **25.** F, fact; **26.** T; **27.** F, observable

Looking Back/Looking Forward

1. Living things are composed of cells, living things maintain homeostasis, living things respond to their external environment

Chapter Test

1. c; **2.** d; **3.** c; **4.** b; **5.** c; **6.** d; **7.** c; **8.** c; **9.** c; **10.** b; **11.** b; **12.** c; **13.** b; **14.** b; **15.** d; **16.** d; **17.** b; **18.** d; **19.** d; **20.** c; **21.** b; **22.** c; **23.** b; **24.** c; **25.** d

Chapter 2

Sections 2.1, 2.2

1. matter; **2.** element; **3.** atom; **4.** proton; **5.** neutron; **6.** electron; **7.** mass; **8.** atomic number; **9.** atomic mass; **10.** isotope; **11.** radioisotope; **12.** molecule; **13.** energy; **14.** potential energy; **15.** kinetic energy; **16.** chemical bond; **17.** ion; **18.** 100;

19. periodic; **20.** atom; **21.** eight; **22.** protons; **23.** two, one; **24.** lose; **25.** covalent; **26.** ionic; **27.** conduct electricity; Completion Exercise: **28.** a) 11, b) 12, c) 11; **29.** a) 11, b) 12, c) 10; **30.** a) 15, b) 31, c) 15; **31.** a) 12, b) 24, c) 12; **32.** a) 1, b) 0, c) 1; **33.** No. Double bonds occur when atoms share two pairs of electrons, and there is no electron sharing in an ionic bond.; **34.** a. Covalent bonds (within each water molecule) and hydrogen bonds (between water molecules). b. Covalent bonds, hydrogen bonds, and ionic bonds between Na+ and Cl−.; **35.** No. Bonds formed between two atoms of the same element are covalent bonds. Both hydrogen atoms are identical, and so they have the same attraction for the shared electrons in the covalent bond. This means the electrons will not spend more time orbiting one nucleus than the other, and the molecule will not be polar.

Sections 2.3, 2.4

1. acid; **2.** base; **3.** buffer; **4.** hydrophilic; **5.** solute; **6.** hydrophobic; **7.** solvent; **8.** pH scale; **9.** acidic; **10.** OH−; **11.** water; **12.** 100; **13.** HCO3−; **14.** H2CO3; **15.** H2CO3; **16.** H+

Sections 2.5, 2.6, 2.7

1. polysaccharide; **2.** cellulose; **3.** lipid; **4.** steroid; **5.** organic molecule; **6.** cholesterol; **7.** dehydration synthesis; **8.** triglyceride; **9.** hydrolysis; **10.** phospholipids; **11.** oligosaccharide; **12.** starch; **13.** monosaccharide; **14.** glycogen; **15.** fatty acid; **16.** macromolecules; **17.** requires; **18.** releases; **19.** carbohydrates, lipids, proteins, nucleic acids; **20.** glucose; **21.** c; **22.** g; **23.** i; **24.** e; **25.** a; **26.** d; **27.** f; **28.** h; **29.** k; **30.** j; **31.** b

Sections 2.8, 2.9, 2.10

1. ADP; **2.** nucleotides; **3.** enzyme; **4.** ATP; **5.** amino acids; **6.** protein; **7.** denaturation; **8.** polypeptide; **9.** DNA, RNA; **10.** catalyst; **11.** F, product; **12.** T; **13.** T; **14.** T; **15.** F, cytosine; **16.** F, ribose; **17.** T; **18.** amino acids; **19.** H, amino group, carboxylic acid group, R group; **20.** when it is longer than 100 amino acids; **21.** primary – sequence of amino acids, secondary – regular coiling due to hydrogen bonding between amino acids, tertiary – complex folding due to various bonds and associations between R groups of amino acids, quaternary – 2 or more polypeptides associated with other in a functional unit; **22.** high temperature, pH changes, changes in the chemical environment; **23.** The physical and chemical environment of the enzyme is critical to maintaining the shape and function of the enzyme. Any deviation from the required environment can affect enzyme shape and function; **23.** *f* b; **24.** h; **25.** c; **26.** f; **27.** a; **28.** g; **29.** d; **30.** e

Looking Back/Looking Forward

1. They are based on carbon, and contain carbon, hydrogen, and oxygen. **2.** Phospholipids have both polar and nonpolar regions; this allows a plasma membrane containing phospholipids to both protect the cell and provide for the cell to function in a polar environment

Chapter Test

1. d; **2.** d; **3.** e; **4.** c; **5.** c; **6.** b; **7.** b; **8.** d; **9.** c; **10.** b; **11.** a; **12.** a; **13.** d; **14.** a; **15.** b; **16.** d; **17.** b; **18.** a; **19.** d; **20.** c; **21.** d; **22.** c; **23.** b; **24.** a; **25.** d

Chapter 3

Sections 3.1, 3.2, 3.3, 3.4

1. diffusion; **2.** cytoplasm; **3.** hypertonic solution; **4.** facilitated transport; **5.** cell doctrine; **6.** active transport; **7.** exocytosis; **8.** plasma membrane; **9.** selective permeability; **10.** sodium-potassium pump; **11.** endocytosis; **12.** receptor protein; **13.** eukaryotes; **14.** isotonic; **15.** osmosis; **16.** passive transport; **17.** prokaryotes; **18.** hypotonic solution; **19.** organelle; **20.** nucleus; **21.** vesicle; **22.** transport protein; **23.** nucleus, cytoplasm, plasma membrane; **24.** microvilli; **25.** transmission electron; **26.** lipid bilayer; **27.** fluid mosaic; **28.** gradient; **29.** Equilibrium; **30.** osmotic pressure; **31.** ADP, P, energy; **32.** extra-cellular; **33.** activity; **34.** Refer to Fig. 3.5 in your text.; **35.** b, c, f; **36.** b, d, f; **37.** b, e, f; **38.** b, d, f, i; **39.** a, e, g; **40.** a,h; **41.** a,h; **42.** small; **43.** opposite to; **44.** cannot; **45.** low, high; **46.** closed, open; **47.** does not; **48.** out of, into; **49.** out of; **50.** Saturated fats have no C to C double bonds and are very compact, while unsaturated fats have C to C double bonds with bends in the hydrocarbon chain. Cholesterol is wedged between fatty acids, preventing saturated fats from becoming too compact and unsaturated fats from spreading out, thereby maintaining the fluidity of the membrane.

Sections 3.5, 3.6

1. mitochondria; **2.** cytoskeleton; **3.** Golgi; **4.** peroxisomes; **5.** ribosomes; **6.** lysosomes; **7.** rough; **8.** flagellum; **9.** centrioles; **10.** nucleus; **11.** vesicle; **12.** smooth; **13.** cilia; **14.** reticulum; **15.** cytosol; **16.** peroxisome; **17.** centrioles; **18.** cytoskeleton; **19.** smooth endoplasmic reticulum; **20.** rough endoplasmic reticulum; **21.** golgi apparatus; **22.** secretory vesicle; **23.** ribosome; **24.** plasma membrane; **25.** nucleus; **26.** mitochondria; **27.** lysosome; **28.** internal surface; as the vesicle fuses with the plasma membrane the internal surface of the vesicle membrane becomes the external surface of the plasma membrane

Section 3.7

1. T; **2.** F, anabolic; **3.** T; **4.** F, product; **5.** F, substrate; **6.** glucose; **7.** cytoplasm; **8.** two PGAL; **9.** two pyruvate; **10.** NADH; **11.** transport of high energy H+ and electrons; **12.** inner compartment of the mitochondria; **13.** two Acetyl CoA; **14.** yes; **15.** inner compartment of the mitochondria; **16.** cyclic; **17.** NAD+ and FAD; **18.** yes; **19.** in the high energy H+ and electrons being carried by NADH and FADH2.; **20.** inner membrane of the mitochondria; **21.** released; **22.** it is an active transport process; **23.** ATP synthase; **24.** diffusion of H+; **25.** oxidative phosphorylation; **26.** the fatty acid tails of a triglyceride molecule contains many hydrogen protons and electrons that can be transported to the electron transport system; **27.** the energy might be released as heat; and the cell would increase metabolism as feedback mechansism indicated that ATP levels were too low

Looking Back/Looking Forward

1. Anabolic, because a larger molecule is being produced from smaller molecules; energy is required because a larger molecule is being built. **2.** active transport

Chapter Test

1. b; **2.** b; **3.** d; **4.** c; **5.** b; **6.** d; **7.** b; **8.** c; **9.** a; **10.** d; **11.** d; **12.** c; **13.** d; **14.** a; **15.** b; **16.** a; **17.** b; **18.** a; **19.** d; **20.** b; **21.** a; **22.** d; **23.** b; **24.** a; **25.** d

Chapter 4

Sections 4.1, 4.2, 4.3

1. cartilage; **2.** adipocytes; **3.** elastic fibers; **4.** adipose tissue; **5.** exocrine gland; **6.** tissue; **7.** blood; **8.** epithelial tissue; **9.** fibroblasts; **10.** endocrine gland; **11.** connective tissue; **12.** glands; **13.** reticular fibers; **14.** collagen fibers; **15.** basement membrane; **16.** bone; **17.** a. simple squamous, b. permits exchange of nutrients, wastes, and gases; **18.** a. simple cuboidal; b. secretes and reabsorbs water and small molecules; **19.** a. simple columnar; b. absorbs nutrients and produces mucus; **20.** a. exocrine gland, b. secretes product into a hollow organ or duct; **21.** a. stratified squamous; b. protection; **22.** a. stratified cuboidal, b. secretes water and ions; **23.** a. stratified columnar, b. secretes mucus; **24.** a. endocrine gland, b. secretes hormones into the bloodstream; **25.** support internal organs, connect structures within the body, store fat, produce blood cells; **26.** living cells, nonliving extra-cellular matrix; **27.** fibrous, specialized; **28.** a. fibers, cells, ground substance, b. collagen fibers (strength), elastic fibers (flexibility without breaking), reticular fibers (form a mesh-like supportive framework); c. fat cells (store fat), mast cells (aid the immune system), white blood cells (aid the immune system), fibroblasts (secrete the fibers of the matrix); **29.** loose (surrounds organs and blood vessels), dense (tendons, ligaments, skin), elastic (surrounds organs that change shape), reticular (provides an internal framework for soft organs); **30.** cartilage, bone, blood, adipose tissue; **31.** a. blood, b. cartilage, c. carti-lage, d. adipose tissue, e. bone, f. bone, g. adipose tissue

Sections 4.4, 4.5

1. glial cells; **2.** smooth muscle tissue; **3.** muscle tissue; **4.** neurons; **5.** skeletal muscle tis-sue; **6.** cardiac muscle tissue; **7.** nervous tissue; **8.** a) skeletal muscle tissue, b) attached to bones, c) multiple; **9.** a) cardiac muscle tissue, b) one, c) involuntary; **10.** a) hollow organs and tubes, b) one, c) involuntary

Section 4.6

1. F, organ; **2.** T; **3.** T; **4.** T; **5.** F, transverse; **6.** F, proximal; **7.** F, diaphragm; **8.** respiratory; **9.** muscular; **10.** circulatory; **11.** reproductive; **12.** urinary; **13.** nervous; **14.** skeletal; **15.** integumentary; **16.** lymphatic; **17.** endocrine; **18.** digestive

Section 4.7

1. blood vessels; **2.** melanocytes; **3.** smooth; **4.** dermicidin; **5.** integumentary; **6.** epidermis; **7.** hypodermis; **8.** sebaceous; **9.** basal; **10.** squamous; **11.** papillae; **12.** keratin; **13.** root; **14.** shaft

Section 4.8

1. interstitial fluid; **2.** homeostasis; **3.** negative feedback; **4.** controlled variable; **5.** sensor; **6.** control center; **7.** effector; **8.** Negative feedback control systems reduce deviations from normal; a positive feedback control system would not be effective in controlling body temperature because any deviation from normal body temperature would be amplified, moving farther away from homeostasis; **9.** The concentration of solutes in the interstitial fluid affect the gradients for both solutes and water, influencing the movements of any substances that move passively.

Looking Back/Looking Forward

1. exocytosis
2. The cells of the upper epidermis are dead and no longer dividing; basal cells and melanocytes are living cells susceptible to damage that can produce the uncontrolled cell division that leads to cancer

Chapter Test

1. c; **2.** b; **3.** d; **4.** b; **5.** c; **6.** d; **7.** d; **8.** d; **9.** b; **10.** b; **11.** d; **12.** d; **13.** d; **14.** c; **15.** c; **16.** a; **17.** b; **18.** c; **19.** d; **20.** c; **21.** b; **22.** a; **23.** d; **24.** c; **25.** d

Chapter 5

Sections 5.1, 5.2, 5.3

1. ligaments; **2.** growth plate; **3.** bone; **4.** chondroblasts; **5.** osteon; **6.** central canal; **7.** cartilage; **8.** bones, ligaments, cartilage; **9.** support, movement, protection, blood cell formation, mineral storage; **10.** a latticework of bone tissue found in spongy bone; **11.** Yellow marrow is located in the shaft of long bones and contains fat cells that provide energy; red marrow is located in spongy bone and contains stem cells that produce blood cells.; **12.** Osteocytes near the central canal of an osteon receive nutrients by diffusion from blood vessels; cell extensions from osteocytes contact each other through canaliculi and pass nutrients through gap junctions.; **13.** hematoma; **14.** fibrocartilage, chondroblasts; **15.** a) fibrocartilage, b) collagen, c) strength and withstands pressure and tension; **16.** a) hyaline, b) collage, c) covers the ends of long bones; **17.** a) elastic, b) flexible, c) outer ear and epiglottis; **18.** epiphysis; **19.** diaphysis; **20.** spongy bone; **21.** compact bone; **22.** yellow bone marrow; **23.** blood vessel; **24.** periosteum; **25.** central cavity; **26.** compact bone; **27.** spongy bone; **28.** osteon; **29.** blood vessels in central canal; **30.** a.7, b.1, c.5, d.4, e.2, f.8, g.6, h.3; **31.** a.7, b.2, c.4, d.1, e.6, f.8, g.5, h.3; **32.** less; **33.** osteoclasts; **34.** high; **35.** one-third, two-thirds; **36.** width; **37.** increase; **38.** increase; **39.** increases; **40.** increases

Sections 5.4, 5.5, 5.6

1. carpals; **2.** metacarpals; **3.** phalanges; **4.** cranium; **5.** maxilla; **6.** mandible; **7.** clavicle; **8.** scapula; **9.** sternum; **10.** ribs; **11.** humerus; **12.** vertebrae; **13.** ulna; **14.** radius; **15.** coxal bone; **16.** sacrum; **17.** femur; **18.** patella; **19.** tibia; **20.** fibula; **21.** tarsals; **22.** metatarsals; **23.** phalanges; **24.** sinusitis; **25.** osteoporosis; **26.** carpal tunnel syndrome; **27.** rheumatoid arthritis; **28.** sprain; **29.** osteoarthritis; **30.** bursitis; Crossword Puzzle: **1.** cervical; **2.** across: symphysis; down: stability; **3.** fibrous; **4.** capsule; **5.** ligament; **6.** tendon; **7.** articulation; **8.** axial; **9.** disk; **10.** across: synovial; down: short; **11.** lumbar; **12.** bursae; **13.** pectoral

Looking Back/Looking Forward

1. Cartilage lacks blood vessels, and so is slow to heal. **2.** Reduced osteoclast activity, and reduced calcium absorption due to low levels of vitamin D, contribute to lack of mineralization of the bone leading to "soft" bones.

Chapter Test

1. d; **2.** b; **3.** c; **4.** d; **5.** a; **6.** a; **7.** b; **8.** c; **9.** a; **10.** e; **11.** c; **12.** c; **13.** b; **14.** e; **15.** b; **16.** b; **17.** b; **18.** b; **19.** d; **20.** c; **21.** d; **22.** d; **23.** c; **24.** c; **25.** a

Chapter 6

Sections 6.1, 6.2

1. sarcomere; **2.** sliding filament mechanism; **3.** fatigue; **4.** skeletal muscle; **5.** neurotransmitter; **6.** sarcoplasmic reticulum; **7.** origin; **8.** oxygen debt; **9.** myofibrils; **10.** myosin; **11.** insertion; **12.** neuromuscular junction; **13.** actin; **14.** motor neuron; **15.** shorten, lengthen; **16.** the same; **17.** insertion, origin; **18.** closer to; **19.** resist; **20.** excitable; **21.** same; **22.** tendon; **23.** multiple nuclei; **24.** myofibrils packed with actin and myosin; **25.** 5; **26.** 2; **27.** 1; **28.** 7; **29.** 6; **30.** 4; **31.** 3; **32.** a; **33.** b; **34.** a; **35.** b; **36.** b; **37.** c; **38.** Transports Ca++ back into the sarcoplasmic reticulum, and allows myosin to detach from actin.; **39.** stored ATP, creatine phosphate, stored glycogen, aerobic metabolism

Section 6.3

1. motor neuron; **2.** motor unit; **3.** muscle tension; **4.** all-or-none; **5.** twitch; **6.** recruitment; **7.** myogram; **8.** summation; **9.** isotonic — maintains a constant force, muscle shortens, skeleton or object is moved; isometric — muscle tension increases, no or little shortening of muscle, no movement of skeleton or object; **10.** size of each motor unit, number of motor units active at one time, how frequently motor units are stimulated; **11.** large motor units generate more force with less control, while small motor units generate less force with more control; **12.** it increases the amount of calcium present; **13.** stimulating a muscle so frequently that it has no chance to relax; **14.** tone; **15.** slow-twitch; **16.** fast-twitch; **17.** aerobic; **18.** strength, endurance; **19.** a) the time between stimulation and the start of contraction, b) nerve impulse travels to the sarcoplasmic reticulum and calcium is released, c) contraction, d) muscle shortens, e) relaxation, f) calcium is transported back into the sarcoplasmic reticulum; **20.** Strength training increases the strength of a particular muscle using resistance; this requires a quick burst of energy, so it targets fast-twitch fibers.; **21.** Running is an aerobic activity requiring the contribution of slow-twitch fibers; slow-twitch fibers do not increase significantly in mass.; **22.** Muscle cells cannot increase in number, however fast-twitch fibers can increase in muscle mass by increasing the number of myofibrils. Strength training, such as weight lifting, can increase muscle mass.

Sections 6.4, 6.5

1. a; **2.** b, c; **3.** b; **4.** a, b; **5.** c; **6.** b, c; **7.** a; **8.** a; **9.** a. inherited genetic disorder resulting in the absence of a membrane protein, b. loss of muscle fibers and muscle wasting; **10.** a. bacterial infection, b. tetanic contractions especially affecting the jaw and neck; **11.** a. ATP depletion, dehydration, ion imbalance, lactic acid accumulation, b. uncontrolled muscle contractions, pain; **12.** a. excessive muscle stretching, resulting in torn fibers, b. internal bleeding swelling, pain; **13.** a. straining or tearing of the fascia around a muscle resulting in inflammation, b. pain

Looking Back/Looking Forward

1. active transport; **2.** Loss of receptors reduces the muscle cell's ability to respond to stimulation by a motor neuron, resulting in muscle weakness

Chapter Test

1. b; **2.** a; **3.** c; **4.** b; **5.** c; **6.** b; **7.** a; **8.** e; **9.** a; **10.** a; **11.** b; **12.** a; **13.** b; **14.** d; **15.** a; **16.** b; **17.** a; **18.** b; **19.** b; **20.** d; **21.** e; **22.** b; **23.** d; **24.** b; **25.** c

Chapter 7

Section 7.1

1. blood; 2. plasma; 3. plasma proteins; 4. albumins; 5. globulins; 6. erythrocytes;
7. leukocytes; 8. hemoglobin; 9. oxyhemoglobin; 10. hematocrit; 11. stem cells;
12. macrophages; 13. erythropoientin; 14. blood doping; 15. heart, blood vessels, blood;
16. transportation, regulation, defense; 17. 5–6 liters, 4–5 liters; 18. red blood cells, white
blood cells, platelets; 19. four protein chains, each containing a heme group with an iron
atom at the center; 20. dark red, venous blood contains more deoxyhemoglobin which
is dark purple, and less oxyhemoglobin; 21. nucleus, organelles; 22. liver, spleen;
23. erythroblasts; 26. thrombopoietin; 25. hematocrit; 26. histamine, basophils; 27. monocytes;
28. Lymphocytes; 29. bilirubin; 30. megakaryocytes; 31. no; 32. temporary; 33. neutral;
34. lower than; 35. detaches from; 36. different sites; 37. shorter; 38. thrombopoietin;
39. White blood cells; 40. myeloblast, granular, first cells to combat infection, they engulf
and destroy foreign cells; 41. myeloblast, granular; defend the body against large
parasites and release chemicals that affect allergic reactions; 42. myeloblast, granular,
secrete histamine; 43. monoblast, agranular, differentiate into macrophages and stimulate
lymphocytes; 44. lymphoblast, agranular, critical to the body's immune system; 45. the
flat rounded red blood cells are flexible enough to squeeze through the smallest blood
vessels, and their flat structure provides for efficient gas exchange; 46. similarities: both
are produced in red bone marrow, both are components of the blood; differences: rbcs lack
a nucleus while wbcs have a nucleus, rbcs transport oxygen and carbon dioxide while wbcs
function in defending the body, rbcs remain in the bloodstream while wbcs migrate out of
the blood and through the tissues; 47. There are no structures in the body capable of
counting the number of cells; 48. Too many red blood cells would be produced, effecting
the thickness of the blood and blood pressure

Section 7.2

1. T; 2. F, blood clot; 3. F, fibrinogen to fibrin; 4. F, blood clotting; 5. T

Section 7.3

1. A antigens, B antibodies, Type A and Type O; 2. B antigens, A antibodies, Type B and
Type O; 3. A and B antigens, no antibodies, all blood types; 4. no antigens, A and B
antibodies, Type O; 5. proteins; 6. antigens; 7. antibodies, lymphocytes; 8. blood, lymph;
9. antigen-antibody complex; 10. antibodies; 11. antibodies, agglutinate; 12. organ damage,
death; 13. transfusion reaction; 14. type A+ also has Rh antigens; 15. antibodies are produced only after exposure; 16. a. Rh antibodies spread from mother to the fetus and
attack fetal red blood cells, b. Mother: Rh–, Father: Rh+, c. no maternal antibodies are
present yet, d. maternal injection of RhoGAM; 17. samples of donor blood are mixed with
recipient plasma and samples of recipient blood are mixed with donor plasma, absence of
agglutination in both combinations

Section 7.4

1. infection of lymphocytes caused by the Epstein-Barr virus; 2. reduced ability of the
blood to transport oxygen due to a variety of causes; 3. uncontrolled division of abnormal
plasma cells in the bone marrow; 4. uncontrolled production of abnormal or immature
white blood cells; 5. proliferation of microorganisms in the blood plasma; 6. a reduction of
the number of platelets in the blood

Looking Back/Looking Forward

1. Diffusion does not require ATP, while active transport does; the use of active transport for movement of oxygen and carbon dioxide into and out of cells would exhaust the body. **2.** No, mature red blood cells lack a nucleus and do not divide.

Chapter Test

1. d; **2.** c; **3.** a; **4.** c; **5.** a; **6.** b; **7.** c; **8.** c; **9.** d; **10.** c; **11.** b; **12.** b; **13.** d; **14.** b; **15.** c; **16.** c; **17.** d; **18.** c; **19.** c; **20.** c; **21.** b; **22.** c; **23.** d; **24.** c; **25.** b

Chapter 8

Sections 8.1, 8.2

1. arteries; **2.** arterioles; **3.** precapillary sphincter; **4.** capillaries; **5.** veins; **6.** atrioventricular valves; **7.** semilunar valves; **8.** coronary arteries; **9.** systole; **10.** diastole; **11.** cardiac cycle; **12.** SA node; **13.** aorta; **14.** arteries, capillaries, veins; **15.** pores, slits; **16.** aneurysm; **17.** vasoconstriction; **18.** Vasodilation; **19.** diffusion; **20.** Capillary beds; **21.** blood pressure; **22.** lymphatic capillaries; **23.** lymph; **24.** veins; **25.** varicose; **26.** pericardium; **27.** pericardial cavity; **28.** arrhythmia; **29.** endocarditis; **30.** pulmonary; **31.** inner layer is the endothelium which is one cell layer thick, middle layer is smooth muscle with elastic fibers, outer layer is tough connective tissue; **32.** blood transport, blood storage, regulation of blood flow into the capillaries; **33.** capillary walls consist of a single layer of endothelium comprised of epithelial cells; **34.** skeletal muscle squeezes the veins, valves prevent blood from flowing backward, pressure associated with breathing pushes blood toward the heart; **35.** sinoatrial node, atrioventricular node, atrioventricular bundle, Purkinje fibers; **36.** epicardium: outermost layer consisting of epithelial and connective tissue, myocardium: thick middle muscular layer responsible for contraction, endocardium: innermost layer of the heart continuous with the lining of the blood vessels; **37.** a) electrical impulses traveling across the atria, b) spread of electrical impulses down the septum and around the two ventricles, c) the end of electrical activity in the ventricles; **38.** a) contract, b) increases, c) open, d) closed; **39.** a) contract, b) closed, c) open, d) pulmonary trunk and aorta; **40.** a) relaxed, b) relaxed, c) close, d) open; **41.** a. right atrioventricular valve, b. right ventricle, c. pulmonary semilunar valve, d. pulmonary trunk, e. lungs, f. pulmonary veins, g. left atrium; h. left atrioventricular valve, i. left ventricle, j. aortic semilunar valve, k. aorta, l. arteries, m. arterioles, n. capillaries, o. venules, p. veins; **42.** Reabsorption would decrease as the solute gradient was decreased.; **43.** Veins have a larger diameter and little elasticity reflecting their role in storing the blood. **44.** jugular vein; **45.** carotid artery; **46.** superior vena cava; **47.** inferior vena cava; **48.** renal vein; **49.** common iliac vein; **50.** common iliac artery; **51.** subclavian vein; **52.** subclavian artery; **53.** aorta; **54.** renal artery; **55.** femoral vein; **56.** femoral artery; **57.** great saphenous vein

Sections 8.3, 8.4, 8.5, 8.6

1. sphygmomanometer; **2.** systolic; **3.** isolated systolic hypertension; **4.** Hypotension; **5.** heart rate, stroke volume; **6.** Sympathetic, medulla oblongata; **7.** one; **8.** angiogram; **9.** CABG; **10.** a) systolic pressure, b) ventricular contraction; **11.** a) diastolic pressure, b) ventricular relaxation; **12.** greater distance from the heart causes blood pressure to dissipate; **13.** blood pressure is steady in the capillaries and pulsatile in the arteries; **14.** blood moves with the pressure gradient; **15.** F, arterial; **16.** F, aorta and carotid arteries;

17. F, decreasing; **18.** T; **19.** T; **20.** a. varied, b. often no symptoms; **21.** a. decreased blood flow to the heart, b. pain and tightness in the chest; **22.** a. death of heart tissue due to oxygen deprivation, b. intense chest pain, nausea, heaviness, difficulty breathing, radiating pain to left arm, back, or abdomen; **23.** a. heart is less efficient at pumping blood due to age, prior heart attacks, deficient valves, hypertension, b. difficulty breathing, swollen legs and ankles, swollen neck veins, weight gain; **24.** a. sudden blockage of a blood vessel by a clot or other obstruction, b. varied; **25.** a. damage to brain cells due to decreased oxygen flow to the brain, b. sudden weakness or paralysis on one side of the body, difficulty speaking, impaired vision, nausea

Looking Back/Looking Forward

1. a) arterial blood pressure, b) baroreceptors, c) cardiovascular center in the brain, d) heart and blood vessels; **2.** Enlargement of the vessels causes endothelial cells to pull apart slightly, increasing their permeability and allowing larger substances to leave the vessel

Chapter Test

1. increases, increasing; **2.** thinner; **3.** larger; **4.** less than; **5.** less than; **6.** increases; **7.** deoxygenated; **8.** d; **9.** c; **10.** a; **11.** b; **12.** c; **13.** c; **14.** c; **15.** a; **16.** a; **17.** a; **18.** a; **19.** b; **20.** c; **21.** d; **22.** a; **23.** d; **24.** c; **25.** c

Chapter 9

Sections 9.1, 9.2

1. F, some; **2.** F, bacterial; **3.** F, prokaryotic; **4.** T; **5.** F, protein coat and genetic material; **6.** F, prions; **7.** T; **8.** F, easily, high; **9.** T; **10.** F, thymosin and thymopoietin; **11.** viruses, bacteria, fungi, protozoa, prions, parasites, worms; **12.** viruses cannot reproduce on their own; **13.** transmissibility, mode of transmission, and virulence; **14.** maintain blood volume, transport absorbed fats and vitamins from the digestive system to the blood, defend the body; **15.** lymph vessels, lymph nodes, thymus gland, tonsils, adenoids, and spleen; **16.** The red pulp is responsible for removing old and damaged red blood cells and breaking down microorganisms. The white pulp contains lymphocytes and helps in defending the body; **17.** a. thymus gland, b. aids maturation of T-cells; **18.** a. lymph nodes, b. filters and cleanses lymph; **19.** a. lymph vessels, b. transports lymph; **20.** a. lymph capillary, b. beginning of lymphatic vessel; **21.** a. adenoids, b. protect the throat; **22.** a. tonsils, b. protect the throat; **23.** a. red pulp, b. contains macrophages that break down microorganisms; **24.** a. white pulp, b. contains lymphocytes that fight infections; **25.** a. spleen, b. removes damaged red blood cells and helps to fight infections; **26.** a) small lymphatic vessels, b) lymph nodes, c) large lymphatic vessels, d) lymphatic ducts, e) subclavian veins; **27.** A lymphatic vessel has spaces between overlapping cells allowing it to absorb large substances, including pathogens.

Sections 9.3, 9.4, 9.5, 9.6, 9.7, 9.8

Crossword Puzzle: **1.** across: cytotoxic; down: cytokines; **2.** presenting; **3.** rejection; **4.** thymus; **5.** primary; **6.** bacteria; **7.** plasma; **8.** antibody; **9.** Monoclonal; **10.** vaccine; **11.** across: major; down: memory; **12.** variable; **13.** perforin; **14.** helper; **15.** lysozyme; **16.** beneficial; **17.** histamine; **18.** interferon; **19.** pyrogens; **20.** antigen; **21.** a) second, first to respond to infection, digest and destroy bacteria, b) second and third, serve a cleanup function, stimulate production of white blood cells, c) second, destroy large parasites and some foreign proteins, d) second, secrete histamine, e) second, destroy tumor cells and virus-infected cells, f) third, destroy abnormal and foreign cells; **22.** the site of injury is exhibiting an inflammatory response with symptoms that reflect defensive actions being taken by the body; **23.** a) 15%, b) enters areas of the body covered by mucus membranes; **24.** a) 0.1%, activates the inflammatory response, b) involved in allergies; **25.** a) 75%, long lived antibodies, b) can cross the placenta; **26.** a) less than 1%, b) function unclear—may activate B cells; **27.** a) 5–10%, b) first antibodies released during an immune response

Sections 9.9, 9.10

1. Allergy; **2.** systemic; **3.** Autoimmune; **4.** lupus erythematosis; **5.** a) few weeks to a few years, b) flu-like symptoms, c) decline then rebound, d) HIV positive; **6.** a) 6 months to 10 years, b) flu-like symptoms, opportunistic infections, and some may show no symptoms, c) declining, d) HIV positive; **7.** a) highly variable, b) opportunistic infections and some cancers, c) less than 200 per cubic millimeter of blood, d) AIDS

Looking Back/Looking Forward

1. No, this would also stop reproduction of human body cells; **2.** See Ch. 14

Chapter Test

1. d; **2.** a; **3.** d; **4.** b; **5.** d; **6.** c; **7.** a; **8.** b; **9.** c; **10.** c; **11.** c; **12.** a; **13.** b; **14.** c; **15.** b; **16.** a; **17.** b; **18.** b; **19.** b; **20.** b; **21.** a; **22.** b; **23.** a; **24.** c; **25.** c

Chapter 10

Sections 10.1, 10.2

1. respiratory; **2.** ventilation; **3.** external respiration; **4.** internal respiration; **5.** cellular respiration; **6.** external nose; **7.** nasal cavity; **8.** cilia; **9.** sinuses; **10.** larynx; **11.** epiglottis; **12.** glottis; **13.** trachea; **14.** breathing, external respiration, internal respiration, cellular respiration; **15.** internal; **16.** cellular; **17.** sound; **18.** nose, pharynx; **19.** larynx, trachea, bronchi, lungs; **20.** nasal septum; **21.** diffusion; **22.** pharynx; **23.** larynx; **24.** Sinuses; **25.** epiglottis, vocal cords; **26.** bronchioles; **27.** fluid, pleural cavity; **28.** three, two; **29.** one, squamous epithelium; **30.** surfactant; **31.** alveoli, pulmonary capillaries

Section 10.3

1. collide; **2.** decreases; **3.** high, low; **4.** pressure gradient; **5.** intercostals, diaphragm; **6.** relaxed; **7.** a) contracts, b) increases, c) increases, d) increases; **8.** decreases; **9.** e) dome shaped, f) downward, g) smaller, h) increases; **10.** i) active, j) passive; **11.** k) active, l) may be passive or active; **12.** the amount of air you can forcibly exhale beyond the tidal volume; **13.** the maximal volume you can inhale after a maximal exhalation; **14.** the amount of air in each breath; **15.** the air remaining in the lungs after forceful exhalation; **16.** the amount of air you can inhale beyond the tidal volume; **17.** inspiratory reserve; **18.** tidal volume; **19.** expiratory reserve; **20.** residual volume; **21.** vital capacity

Section 10.4

1. a. 104, b. 100, c. into the pulmonary capillaries, d. 40, e. 46, f. into the alveoli, g. external; **2.** a. 100, b. <40, c. into the cell, d. 46, e. >46, f. into the capillary, g. internal; **3.** a. hemoglobin, oxygen, oxyhemoglobin, b. reversible, c. neutral pH, cool temperature, high PO2 in the lungs; **4.** a. hemoglobin, carbon dioxide, carbaminohemoglobin, b. about 20%; **5.** a. bicarbonate, b. about 70%

Sections 10.5, 10.6

1. nervous; **2.** 4–5; **3.** respiratory; **4.** medulla oblongata; **5.** intercostals; **6.** expands; **7.** downward; **8.** inhale; **9.** stretch; **10.** respiratory; **11.** exhalation; **12.** relax; **13.** upward; **14.** exhale; **15.** CO_2; **16.** H+; **17.** O_2; **18.** CO_2; **19.** H+; **20.** H+; **21.** cerebrospinal; **22.** deeper and more frequent; **23.** CO_2; **24.** 2–3; **25.** O_2; **26.** carotid and aortic; **27.** carotid arteries and aorta; **28.** respiratory; **29.** increase; **30.** 20%; **31.** Cells in the medulla oblongata control inspiration, expiration, and rate of breathing; damage to these cells can disrupt breathing; **32.** blood buffers maintain a constant pH in the blood, preventing the acidic conditions that are detected by cells monitoring the cerebrospinal fluid; **33.** a. hyperactive immune system causing bronchial muscle contraction and swelling, b. shortness of breath, chest tightness; **34.** a. damage to the alveoli resulting in reduced elasticity, due to either smoking or inheritance, b. difficulty breathing; **35.** a. inflammation of the bronchi, may be due to bacterial infection, b. breathlessness and coughing; **36.** a. lung inflammation due to bacterial or viral infection, b. fever, chills, shortness of breath, cough, pain; **37.** a. bacterial infection beginning in the lungs, may spread to the bloodstream, b. coughing, chest pain, shortness of breath, fever, weight loss; **38.** a. consumption of food contaminated by the bacterium Clostridium botulinum, b. difficulty swallowing, vomiting, paralysis of respiratory muscles; **39.** a. usually caused by smoking, may also result from exposure to radon gas or asbestos, b. chest pain, bloody cough; **40.** a. genetic defect resulting in production of thick sticky mucus in the lungs, b. difficulty breathing, respiratory infections **41.** a. failure of the left side of the heart, b. fluid accumulation in the lungs; **42.** a. collapse of one or more lobes of the lungs, b. difficulty breathing

Looking Back/Looking Forward

1. Macrophages can defend the body against pathogens and foreign substances in the air.; **2.** vital capacity decreases, residual volume increases

Chapter Test

1. d; **2.** b; **3.** e; **4.** b; **5.** a; **6.** d; **7.** d; **8.** a; **9.** c; **10.** c; **11.** a; **12.** a; **13.** c; **14.** e; **15.** b; **16.** d; **17.** b; **18.** d; **19.** d; **20.** d; **21.** a; **22.** c; **23.** c; **24.** a; **25.** c

Chapter 11

Sections 11.1, 11.2, 11.3, 11.4, 11.5

1. nervous; **2.** neurons; **3.** cell body; **4.** dendrite; **5.** axon; **6.** axon hillock; **7.** depolarization; **8.** hyperpolarization; **9.** graded potential; **10.** summation; **11.** neuroglial cells; **12.** Schwann cells; **13.** saltatory conduction; **14.** oligodendrocytes; **15.** synapse; **16.** synaptic cleft; **17.** Central Nervous System (CNS)—brain and spinal cord, Peripheral Nervous System (PNS)—all nerves and components that lie outside the CNS; **18.** sensory division—carries information to the brain and spinal cord, motor division—carries information away from the brain; **19.** somatic division—controls skeletal muscle, autonomic division—controls smooth muscle, cardiac muscle, and glands; **20.** sympathetic and parasympathetic, they are antagonistic; **21.** sensory—respond to stimuli by transmitting information to the CNS, interneurons—transmit impulses between components of the CNS, motor—carry information away from the CNS; **22.** the sodium-potassium pump helps to create a negatively charged environment inside a cell by transporting three Na^+ out of the cell for every two K^+ transported into the cell; this results in the net loss of one positive charge from the cell's internal environment; **23.** saves energy, speeds transmission, and aids regeneration of damaged neurons Diagram: refer to your text; **24.** resting membrane potential: the normal membrane potential at rest; **25.** depolarizing graded potential: membrane voltage moves closer to zero; **26.** hyperpolarizing graded potential: membrane voltage becomes more negative; **27.** summation: the addition of graded potentials in space or time; **28.** action potential: a sudden, temporary reversal of the voltage difference across the cell membrane; **29.** threshold: membrane voltage required to trigger an action potential; **30.** are not; **31.** into; **32.** is; **33.** increases; **34.** open, into; **35.** short-lived; **36.** do not, one; **37.** depolarize, postsynaptic; **38.** presynaptic; **39.** chemically sensitive; **40.** postsynaptic; **41.** a, d, g, i, j, b, e, c, h, f; **42.** If sodium is allowed to accumulate inside the cell body, the sodium gradient will reach equilibrium; if this occurs there will be no gradient to drive the movement of sodium into the cell the next time sodium channels open and no message could be sent by the neuron.; **43.** The postsynaptic neuron receives information from many presynaptic neurons and must integrate all of the messages received; the message received from a specific presynaptic neuron may not result in a message being sent.

Sections 11.6, 11.7

Crossword Puzzle: **1.** (across) postganglionic; **1.** (down) parasympathetic; **2.** preganglionic; **3.** sympathetic; **4.** autonomic; **5.** two; **6.** somatic; **7.** meninges; **8.** white; **9.** ventral; **10.** ventricle; **11.** ganglia; **12.** nerves; **13.** Cranial; **14.** dorsal; **15.** gray; **16.** reflex; **17.** Spinal

Sections 11.8, 11.9, 11.10, 11.11

1. thalamus; **2.** forebrain; **3.** cerebellum; **4.** Hypothalamus; **5.** pons; **6.** midbrain; **7.** Medulla oblongata; **8.** corpus callosum; **9.** hindbrain; **10.** cerebrum; **11.** medulla oblongata; **12.** medulla oblongata; **13.** reticular formation; **14.** hypothalamus, thalamus, limbic system, cerebrum; **15.** limbic system; **16.** frontal

Sections 11.12, 11.13

1. a drug that affects higher brain functions; **2.** because they cross the blood-brain barrier; **3.** they influence the concentration or mode of action of brain neurotransmitters; **4.** cocaine prolongs the presence of dopamine in the synaptic cleft, causing repeated stimulation of a neuron; **5.** the liver detoxifies drugs; rising levels of liver enzymes necessitate higher dosages to achieve the same effect; **6.** the need to continue obtaining and using a substance despite one's better judgment and good intentions; **7.** concussion; **8.** spinal cord injury; **9.** encephalitis; **10.** meningitis; **11.** rabies; **12.** epilepsy; **13.** Parkinson's disease; **14.** Alzheimer's disease; **15.** brain tumor

Looking Back/Looking Forward

1. The thick lipid layer created when a Schwann cell wraps around the axon is nonpolar and nonconductive; **2.** Regeneration of the CNS neurons may lead to alteration of critical pathways in the CNS

Chapter Test

1. c; **2.** a; **3.** b; **4.** c; **5.** d; **6.** c; **7.** b; **8.** a; **9.** c; **10.** d; **11.** a; **12.** d; **13.**a; **14.**d; **15.**b; **16.**b; **17.** d; **18.** b; **19.** d; **20.** b; **21.** d; **22.** a; **23.** c; **24.** b; **25.** c

Chapter 12

Sections 12.1, 12.2, 12.3

1. stimulus mechanoreceptor; **2.** receptor; **3.** muscle spindles; **4.** mechanoreceptor; **5.** chemoreceptor; **6.** tastants; **7.** umami; **8.** olfactory; **9.** pain, light pressure, temperature changes; **10.** light touch and pressure; **11.** deep pressure and frequent vibration; **12.** beginning and end of light pressure; **13.** ongoing pressure; **14.** a sensation is conscious awareness of a stimulus while perception is understanding the meaning of the stimulus; **15.** mechanoreceptors—mechanical energy, thermoreceptors—heat or cold, pain receptors—tissue damage and excessive pressure or temperature, chemoreceptors—chemicals, photoreceptors—light; **16.** receptors that function in negative feedback loops without conscious awareness; **17.** temperature, touch, vibration, pressure, pain, awareness of body position; **18.** primary somatosensory area of the parietal lobe, primary motor area of the frontal lobe; **19.** sweet—tip, sour—sides, salty—tip and sides, bitter—back; **20.** smell receptors are located on sensory neurons, and there are only four tastant classes while there are over 1000 odorant chemicals; **21.** Both smell and pain provide important signals for survival.

Sections 12.4, 12.5

1. outer ear; **2.** middle ear; **3.** inner ear; **4.** pinna; **5.** malleus; **6.** incus; **7.** auditory canal; **8.** tympanic membrane; **9.** semicircular canals; **10.** vestibular nerve; **11.** cochlear nerve; **12.** vestibule; **13.** oval window; **14.** stapes; **15.** cochlea; **16.** round window; **17.** auditory tube; **18.** T; **19.** T; **20.** F, frequency; **21.** F, pinna; **22.** F, outer; **23.** F, tympanic membrane; **24.** F, oval window; **25.** F, throat; **26.** F, cochlear; **27.** T; **28.** F, hair cells; **29.** F, basilar membrane; **30.** T; **31.** F, base; **32.** T; **33.** T; **34.** F, vestibule; **35.** T

Sections 12.6, 12.7

1. macula; **2.** fovea centralis; **3.** optic nerve; **4.** optic disc; **5.** Sclera; **6.** Canal of Schlemm; **7.** Cornea; **8.** Aqueous humor; **9.** Iris; **10.** Pupil; **11.** Lens; **12.** Vitreous humor; **13.** Ciliary muscle; **14.** Choroid; **15.** Retina; **16.** e; **17.** k; **18.** c; **19.** m; **20.** d; **21.** n; **22.** g; **23.** a; **24.** i; **25.** l; **26.** b; **27.** f; **28.** j; **29.** h; **30.** photopigment; **31.** Rhodopsin; **32.** Cones; **33.** cornea, lens; **34.** ganglion cells; **35.** myopia, long; **36.** retinal detachment; **37.** glaucoma; **38.** cataracts; **39.** otitis media; **40.** Ménière's syndrome; **41.** myopia

Looking Back/Looking Forward

1. Binding of these receptors interferes with the processes that transmit pain impulses regardless of the type of molecule that binds; **2.** reduced ability to taste sweet substances

Chapter Test

1. c; **2.** a; **3.** b; **4.** d; **5.** a; **6.** a; **7.** a; **8.** c; **9.** b; **10.** c; **11.** d; **12.** c; **13.** d; **14.** c; **15.** a; **16.** a; **17.** a; **18.** c; **19.** c; **20.** a; **21.** b; **22.** a; **23.** c; **24.** c; **25.** b

Chapter 13

Sections 13.1, 13.2, 13.3

1. endocrine system; **2.** hormones; **3.** endocrine glands; **4.** target cells; **5.** steroid hormones; **6.** nonsteroid hormones; **7.** second messenger; **8.** hypothalamus; **9.** pituitary gland; **10.** neuroendocrine cells; **11.** bloodstream, cells; **12.** 50; **13.** receptors; **14.** endocrine, nervous;

15. steroid; **16.** nonsteroid; **17.** steroid; **18.** nonsteroid; **19.** target cells; **20.** pituitary gland, internal environment, temperature, carbohydrate metabolism; **21.** hypothalamus, posterior pituitary; **22.** neural, endocrine; **23.** ADH; **24.** c, i; **25.** e, k; **26.** d, j; **27.** a, p; **28.** h, m; **29.** b, o; **30.** g, l; **31.** f, n; **32.** endocrine hormones reach almost all cells, a hormone acts only on its target cell, endocrine system messages are slower than nervous system messages, and the endocrine system interacts with the nervous system; **33.** a. cholesterol is used in the synthesis of steroid hormones, while nonsteroid hormones utilize amino acids, b. steroid hormones are lipid soluble, while nonsteroid hormones are not, c. steroid hormones exert an effect after they enter a target cell, while nonsteroid hormones bind to the target cell surface and trigger a series of events that affect the target cell; **34.** Steroid hormones act by affecting DNA activity and increasing or decreasing production of a critical protein; because nonsteroid hormones affect proteins that have already been synthesized, their effect is faster; **35.** hormones reach almost every cell in the body, and endocrine messages are a much slower form of communication than messages sent by the nervous system.

Sections 13.4, 13.5, 13.6, 13.7, 13.8

1–14. refer to Tables 13.2 and 13.3 in text; (Crossword Puzzle: **1.** Parathyroid; **2.** aldosterone; **3.** calcitonin; **4.** glucagons; **5.** epinephrine; **6.** insulin; **7.** cortisol; **8.** thyroxin; **9.** Atrial; **10.** TSH;) **11.** Insulin promotes the uptake of glucose by cells, while cortisol promotes an increase in blood glucose levels. The administration of cortisol-like drugs to a patient who also has diabetes can make the diabetic condition much more difficult to control.

Sections 13.9, 13.10

1. increases mucus secretion, dilates blood vessels, and increases the leakiness of capillaries; **2.** locally controls blood flow, contributes to inflammatory response, aids blood clotting; **3.** regulates smooth muscle contraction in the digestive tract, controls penile erection; **4.** influences cell division and the direction of cell growth; **5.** a. varied, b. slow growth in children, adults may have swelling, lethargy, and weight gain; **6.** a. may be an autoimmune disorder, b. includes protruding eyes; **7.** a. decreased secretion of cortisol and aldosterone by the adrenal cortex, b. fatigue, weakness, abdominal pain, weight loss, bronzed skin color; **8.** a. excessive cortisol production, may be due to a tumor, b. muscle weakness and fatigue, swelling, high blood pressure; **9.** a. pancreas cannot produce enough insulin, may be caused by a virus, b. dehydration, thirst, fatigue, frequent urination, blurred vision, slow wound healing; **10.** a. cells are unable to respond to insulin, caused by genetics and lifestyle; b. symptoms similar to those for type I diabetes

Looking Back/Looking Forward

1. The endocrine system is critical in maintaining homeostasis of many body processes, positive feedback amplifies a deviation from homeostasis while negative feedback reduces a deviation from normal; **2.** severe dehydration

Chapter Test

1. d; **2.** a; **3.** d; **4.** b; **5.** c; **6.** b; **7.** b; **8.** b; **9.** b; **10.** d; **11.** d; **12.** b; **13.** f; **14.** d; **15.** c; **16.** b; **17.** a; **18.** b; **19.** b; **20.** b; **21.** c; **22.** b; **23.** d; **24.** c; **25.** c

Chapter 14

Sections 14.1, 14.2, 14.3, 14.4, 14.5

1. digestive; **2.** peristalsis; **3.** segmentation; **4.** crown; **5.** epiglottis; **6.** lower esophageal sphincter; **7.** chyme; **8.** duodenum; **9.** jejunum and ileum; **10.** salivary glands, f; **11.** liver, a; **12.** gallbladder, l; **13.** pancreas, d; **14.** appendix, m; **15.** mouth, n; **16.** pharynx, c; **17.** esophagus, g; **18.** stomach, j; **19.** small intestine, k; **20.** large intestine, h; **21.** sigmoid colon, n; **22.** rectum, i; **23.** anus, b; **24.** salivary glands, liver, gallbladder, pancreas; **25.** mechanical processing and movement, secretion, digestion, absorption, elimination; **26.** mucin—holds food particles together, amylase—digests carbohydrates, bicarbonate—maintains pH, lysozyme—inhibits bacterial growth; **27.** food storage, digestion, regulation of delivery; **28.** openings to gastric glands, the secretions found in gastric juice are HCl, pepsinogen, and fluid; **29.** digestion and absorption; **30.** longitudinal layer of smooth muscle; **31.** circular layer of smooth muscle; **32.** lumen; **33.** mucosa; **34.** submucosa; **35.** muscularis; **36.** serosa; **37.** muscularis; **38.** submucosa; **39.** serosa; **40.** muscularis

Sections 14.6, 14.7, 14.8, 14.9

1. m, o; **2.** b, d, e, h, i. l; **3.** c, j, n; **4.** k; **5.** a; **6.** f; **7.** g; **8.** a) stomach, b) small intestine, c) pepsin, proteases, d) jejunum, ileum, e) active transport; **9.** a) mouth, b) small intestine, c) amylase, d) jejunum, ileum, e) active transport; **10.** a) small intestine, b) small intestine, c) lipase, d) jejunum, ileum, e) exocytosis; **11.** F, monoglycerides; **12.** F, fatty acids and monoglycerides; **13.** T; **14.** F, increases; **15.** F, secretin; **16.** T

Sections 14.10, 14.11, 14.12, 14.13

1. Complex carbohydrates consist of many sugar units linked together, while simple sugars are much smaller. Complex carbohydrates release these sugar units slowly, and also provide additional healthy components.; **2.** Saturated fats are found primarily in meat and dairy products; they are unhealthy because they increase LDLs in the blood and contribute to heart disease.; **3.** Essential amino acids are those that the body cannot manufacture and must be obtained through the diet.; **4.** A complete protein contains all 20 amino acids; most animal proteins are complete protein.; **5.** carbohydrates: 45–65%, lipids: 20–35%, proteins: 15%; **6.** Fat soluble vitamins are stored in the body's fat tissues and released over time, so they don't need to be consumed daily.; **7.** Fiber is a material found in vegetables, fruits, and grains; it does not provide energy because we cannot break it down.; **8.** a. higher in males; b. increase; c. increase; d. decrease; e. decrease; **9.** loss of the enzyme that digest the mile sugar lactose; **10.** diarrhea, gas, bloating, cramps; **11.** Diverticulosis; **12.** Hepatitis A; **13.** genetic; **14.** fatigue and abdominal pain; **15.** inadequate fiber may contribute; **16.** may have no symptoms; **17.** Gallstones; **18.** may have no symptoms, may cause severe pain; **19.** Anorexia nervosa; **20.** extra weight loss, fatigue, hair loss, eventual osteoporosis; **21.** unknown, symptoms caused by cycle of eating and intentional vomiting; **22.** ulcers, heartburn, rectal bleeding, gum and tooth damage

Looking Back/Looking Forward

1. The increased solute concentration creates a hypertonic environment for intestinal cells, causing them to lose excessive amounts of water into the lumen of the intestines; **2.** Estrogen is stored in fat cells, loss of fat cells in anorexia results in lower levels of estrogen

Chapter Test

1. c; **2.** b; **3.** b; **4.** c; **5.** a; **6.** d; **7.** b; **8.** b; **9.** a; **10.** c; **11.** a; **12.** d; **13.** d; **14.** d; **15.** a; **16.** d; **17.** a; **18.** c; **19.** c; **20.** c; **21.** c; **22.** d; **23.** c; **24.** d; **25.** a

Chapter 15

Sections 15.1, 15.2, 15.3

1. excretion; **2.** urea; **3.** ureter; **4.** urethra; **5.** nephron; **6.** glomerulus; **7.** afferent arterioles; **8.** efferent arterioles; **9.** peritubular capillaries; **10.** vasa recta; **11.** renal pelvis; **12.** urinary bladder; **13.** ureters, bladder, urethra; **14.** by peristalsis; **15.** proximal tubule, loop of Henle, distal tubule, collecting duct; **16.** removal of water, ions, and nutrients which will be reabsorbed by the proximal and distal tubules; **17.** plasma fluid is filtered out of the glomerulus; **18.** it is shorter in females; **19.** efferent arterioles; **20.** glomerulus; **21.** peritubular capillaries; **22.** descending limb; **23.** ascending limb; **24.** loop of Henle; **25.** collecting duct; **26.** distal tubule; **27.** afferent arteriole; **28.** glomerular capsule; **29.** proximal tubule

Sections 15.4, 15.5, 15.6

1. podocytes; **2.** blood pressure; **3.** water, solutes, protein; **4.** aquaporins; **5.** H^+, NH_4^+; **6.** K^+; **7.** concentration, renal medulla; **8.** urea; **9.** countercurrent; **10.** micturition reflex; **11.** smooth, skeletal; **12.** glomerulus filtration, tubular reabsorption, tubular secretion; **13.** located on the outer surface of glomerular capillaries, specialized for filtration; **14.** glucose, amino acids, bicarbonate, water, sodium; **15.** substances move from the capillaries into the tubules to be secreted, active transport and diffusion; **16.** low blood volume; **17.** increased permeability of the collecting duct to water; **18.** the glomerular capillaries do not normally allow proteins to pass into the filtrate; proteins in the urine can indicate damage to the glomerular capillaries; **19.** c, d; **20.** f; **21.** d, e, f; **22.** b; **23.** a; **24.** a, b, f

Sections 15.7, 15.8

1. blood, increases; **2.** increases, concentrated; **3.** hypothalamus, decrease; **4.** adrenal gland; **5.** increases, sodium; **6.** increase; **7.** increased, afferent; **8.** angiotensinogen; **9.** aldosterone; **10.** heart, increased; **11.** T; **12.** T; **13.** T; **14.** T; **15.** T; **16.** T; **17.** F, decrease; **18.** F, calcium and phosphorus; **19.** T; **20.** T; **21.** F, PTH; **22.** crystallization of minerals in the urine, they often require no treatment but can be removed surgically or crushed when necessary; **23.** bacteria, because the urethra is shorter in women; **24.** acute renal failure is short-term and often correctable, chronic renal failure is long-term and irreversible; **25.** benefits include at-home use and freedom of movement, infection is a risk

Looking Back/Looking Forward

1. a. They enter the reactions of cellular respiration to produce ATP, b. Catabolic; **2.** As the kidneys concentrate urine, they also concentrate any toxins in the urine, increasing the health risks of the toxins

Chapter Test

1. b; **2.** d; **3.** c; **4.** c; **5.** a; **6.** a; **7.** c; **8.** d; **9.** b; **10.** a; **11.** c; **12.** c; **13.** c; **14.** b; **15.** b; **16.** b; **17.** a; **18.** c; **19.** a; **20.** b; **21.** a; **22.** b; **23.** d; **24.** b; **25.** a

Chapter 16

Section 16.1

1. bladder; **2.** pubic symphysis; **3.** vas deferens; **4.** penis; **5.** erectile tissue; **6.** testis; **7.** vertebral column; **8.** ureter; **9.** rectum; **10.** seminal vesicle; **11.** ejaculatory duct; **12.** prostate gland; **13.** bulbourethral gland; **14.** epididymis; **15.** vas deferens; **16.** seminal vesicle; **17.** prostate gland; **18.** bulbourethral gland; **19.** urethra; **20.** epididymis; **21.** testis; **22.** penis; **23.** seminiferous tubule; **24.** gamete; **25.** 46, 23; **26.** inhibin; **27.** negative, pituitary gland; **28.** GnRH; **29.** LH; **30.** scrotum; **31.** seminal vesicles; **32.** semen; **33.** puberty, every day; **34.** Sertoli; **35.** testis; **36.** scrotum; **37.** seminiferous tubules; **38.** epididymis; **39.** vas deferens; **40.** ejaculatory duct; **41.** urethra; **42.** seminal vesicles; **43.** prostate gland; **44.** bulbourethral gland; **45.** spermatogonium; **46.** Sertoli cell; **47.** primary spermatocyte; **48.** secondary spermatocyte; **49.** early spermatids; **50.** late spermatids; **51.** immature sperm; **52.** acrosome; **53.** head; **54.** midpiece; **55.** tail; **56.** diploid; **57.** haploid; **58.** mitosis; **59.** secondary spermatocyte; **60.** spermatid; **61.** no; **62.** DNA; **63.** enzymes that aid penetration of the egg; **64.** to house mitochondria that provide energy for sperm movement; **65.** movement of sperm

Section 16.2

1. ovaries; **2.** ovaries; **3.** oviduct; **4.** fimbriae; **5.** uterus; **6.** cervix; **7.** endometrium; **8.** vagina; **9.** vulva; **10.** clitoris; **11.** mammary glands; **12.** 1-2; **13.** upper one third of the oviduct; **14.** the endometrium is the inner layer of epithelial tissue, glands, connective tissue, and blood where a fertilized egg implants, the myometrium is the outer layer of smooth muscle that helps to expel the fetus during labor

Sections 16.3, 16.4, 16.5

1. ovarian; **2.** primary; **3.** Meiosis I; **4.** puberty; **5.** Meiosis II; **6.** granulose; **7.** follicle; **8.** FSH; **9.** anterior; **10.** pituitary; **11.** GnRH; **12.** hypothalamus; **13.** ovulation; **14.** glycoprotein; **15.** zona; **16.** pellucida; **17.** antrum; **18.** progesterone; **19.** Meiosis I; **20.** secondary; **21.** Graafian; **22.** LH; **23.** ovulation; **24.** corpus; **25.** luteum; **26.** endometrium; **27.** decrease; **28.** ovarian; **29.** days 1–5; **30.** estrogen and progesterone levels decline; **31.** days 6–14; **32.** rising levels of estrogen and progesterone; **33.** ovulation; **34.** secretory phase; **35.** progesterone; **36.** implantation of a fertilized egg; **37.** degeneration of the corpus luteum; **38.** before; **39.** rising levels of estrogen in the first half of the proliferative phase trigger a surge in LH; **40.** high levels progesterone and estrogen, secreted by the corpus luteum, feed back to the hypothalamus and inhibit FSH secretion; **41.** vasectomy; **42.** Depo-Provera; **43.** spermicide; **44.** intrauterine device (IUD); **45.** rhythm method; **46.** Lunelle™; **47.** Ortho Evra®; **48.** Mifeprex; **49.** Structurally, the sperm needs only to transport the male chromosomes, while the egg must contain sufficient cytoplasm to support rapid cell division after fertilization. Functionally, many sperm are required to successfully fertilize one egg – many sperm will never reach the egg, and those reaching the egg must release enzymes from their acrosomes that facilitate one final sperm succeeding in entering the egg.

Sections 16.6, 16.7

1. failure to achieve pregnancy after one year of trying; **2.** 60 million sperm per ejaculate; **3.** PID; **4.** eggs are extracted from the female, fertilized outside the body, and embryos are implanted into the uterus; **5.** in GIFT fertilization occurs within the body when unfertilized eggs and sperm are surgically placed into a fallopian tube; **6.** a. bacterial infection, b. symptoms include discharge and painful urination, infected individuals may have no symptoms, c. medication; **7.** a. bacterial infection, b. occurs in three stages, refer to text for symptoms; c. medication; **8.** a. bacterial infection, b. discharge, painful urination, swollen lymph nodes, PID, c. medication; **9.** viral infection, early flu like symptoms with opportunistic infections in later stages, drugs may keep in remission; **10.** viral infection, inflammation of the liver, vaccination may prevent infection; **11.** a. viral infection, b. blisters, painful urination, swollen lymph nodes in groin, fever, c. no cure available, suppressive drugs only; **12.** a. viral infection, b. warts in genital area, c. no cure available, can be removed by surgery or freezing but tend to reappear

Looking Back/Looking Forward

1. The female immune system has never been exposed to sperm and would normally attack sperm as foreign; **2.** Development cannot continue and the indeterminate structures will develop as a female in the absence of testosterone

Chapter Test

1. d; **2.** d; **3.** a; **4.** c; **5.** d; **6.** c; **7.** a; **8.** c; **9.** b; **10.** a; **11.** c; **12.** c; **13.** d; **14.** d; **15.** c; **16.** c; **17.** b; **18.** a; **19.** a; **20.** c; **21.** c; **22.** c; **23.** d; **24.** d; **25.** d

Chapter 17

Sections 17.1, 17.2

1. not labeled; **2.** d, g; **3.** m; **4.** b, f, g; **5.** n; **6.** c, e, f, g, i; **7.** p; **8.** f, g, k; **9.** l; **10.** a, h; **11.** o; **12.** f, g, j; **13.** mitotic phase; **14.** cytokinesis; **15.** chromosomes; **16.** histones; **17.** sister chromatids, centromere; **18.** gene; **19.** uncoiling; **20.** DNA polymerase; **21.** replication bubbles; **22.** mutations; **23.** repair enzymes; **24.** promoter; **25.** polymerase; **26.** codon; **27.** tRNA, mRNA; **28.** to keep replication from taking too long; **29.** mistakes in DNA replication, chemicals, physical forces; **30.** the primary transcript contains introns and exons while the mRNA molecule contains only exons; **31.** mRNA carries the instructions on the correct sequence of amino acids in the protein, rRNA complexes with protein to become ribosomal subunits, tRNA transports amino acids from the cytoplasm to the ribosome during protein synthesis; **32.** a. replication, b. nucleus; **33.** a. transcription, b. nucleus; **34.** a. translation, b. cytoplasm; **35.** TGATGGGTACG-GTC; **36.** UGAUGGGUACGGUC; **37.** met-gly-thr-val; **Initiation.** e, c, b; **Elongation.** f, a, h; **Termination.** d, i, g

Sections 17.3, 17.4, 17.5

1. anaphase; **2.** diploid cell; **3.** haploid cell; **4.** crossing-over; **5.** mitotic spindle; **6.** homologous chromosomes; **7.** prophase; **8.** telophase; **9.** metaphase; **10.** Meiosis I; **11.** Meiosis I; **12.** Mitosis; **13.** Meiosis II; **14.** Mitosis; **15.** Meiosis I; **16.** Meiosis II; **17.** 46; **18.** a prophase cell has twice the amount of DNA because the chromosomes are replicated; each chromosome consists of two pieces of DNA; **19.** yes; **20.** chromatin; **21.** crossing over in Prophase I, aligning of the chromosome pairs in Metaphase I, and fertilization; **22.** F, different; **23.** T; **24.** T; **25.** T; **26.** T; **27.** T; **28.** T

Sections 17.6, 17.7

1. b; **2.** b; **3.** a; **4.** b; **5.** a; **6.** a; **7.** b; **8.** a, b;

Looking Back/Looking Forward

1. Mitosis occurs normally, but is not followed by cytokinesis, producing a single daughter cell with two nuclei; **2.** Homologous chromosomes may fail to separate in Anaphase I, or sister chromatids may fail to separate in Anaphase II.

Chapter Test

1. a; **2.** d; **3.** a; **4.** c; **5.** d; **6.** c; **7.** c; **8.** c; **9.** b; **10.** b; **11.** c; **12.** a; **13.** b; **14.** c; **15.** b; **16.** f; **17.** d; **18.** c; **19.** c; **20.** d; **21.** a; **22.** c; **23.** d; **24.** d; **25.** d

Chapter 18

Sections 18.1, 18.2, 18.3

1. hyperplasia; **2.** tumor; **3.** benign; **4.** dysplasia; **5.** metastasis; **6.** structural; **7.** regulatory; **8.** growth factors; **9.** oncogene; **10.** p53; **11.** mutator; **12.** age; **13.** carcinogenesis; **14.** carcinogen; **15.** loss of control over cell division; **16.** when they crowd normal cells; **17.** nucleus enlarges, volume of cytoplasm decreases, loss of specializations; **18.** when some cells lose all organization, structure, and regulatory control; **19.** malignant tumors metastasize; **20.** 1 in 3, 1 in 4; **21.** cells must divide uncontrollably and break away from surrounding cells; **22.** proto-oncogenes and tumor suppressor genes; **23.** see section 18.3; **24.** UV B rays in sunlight; **25.** tobacco and diet; **26.** e; **27.** c, j; **28.** a,h; **29.** f, g; **30.** b, i; **31.** d; **32.** Every time a cell divides there is an opportunity for a mutation to occur. During pregnancy there is no ovulation and so no cell division needed to repair the area of the ruptured follicle in the ovary. **33.** Viruses insert their DNA into host cell DNA and may disrupt normal host cell gene function.

Sections 18.4, 18.5

1. PET scan; **2.** genetic testing; **3.** enzyme testing; **4.** chemotherapy; **5.** photodynamic therapy; **6.** gene therapy; **7.** "starving" cancer; **8.** cancer cells divide rapidly, while most normal body cells do not undergo continuous cell division; **9.** the cancer antigen is used to produce antibodies that will target the cancer cell for destruction by the immune system

Looking Back/Looking Forward

1. S phase; **2.** no

Chapter Test

1. d; **2.** c; **3.** b; **4.** d; **5.** b; **6.** c; **7.** d; **8.** b; **9.** c; **10.** c; **11.** c; **12.** a; **13.** b; **14.** d; **15.** c; **16.** c; **17.** c; **18.** c; **19.** a; **20.** b; **21.** d; **22.** b; **23.** a; **24.** a; **25.** d

Chapter 19

Sections 19.1, 19.2, 19.3, 19.4

1. gene; **2.** genome; **3.** genetics; **4.** autosomes; **5.** homologous chromosomes; **6.** locus; **7.** allele; **8.** homozygous; **9.** heterozygous; **10.** mutation; **11.** genotype; **12.** phenotype; **13.** 22, 1; **14.** 2; **15.** during gamete formation each gamete receives only one allele of each gene alleles of different genes are distributed to egg and sperm cells independently of each other, no; **16.** upper case letter, lower case letter, when no dominant alleles are present; **17.** in codominance a heterozygote will express both alleles equally, while in incomplete dominance a heterozygote will show a phenotype intermediate between the dominant and recessive trait; **18.** eye color, height, skin pigment; **19.** when they are located very close together, crossing over in meiosis; **20.** Aa; **21.** Aa; **22.** AA; **23.** Aa; **24.** Aa; **25.** aa; **26.** a. aa, b. normal pigmentation, c. Aa, d. normal pigmentation, e. 50%;

	a	a
A	Aa	Aa
a	aa	aa

27. a. incomplete dominance,
b.

	W	W
w	Ww	Ww
w	Ww	Ww

c.

	AW	Aw	aW	aw
aW	AaWW	AaWw	aaWW	aaWw

28. a. HbAHbS, b. HbSHbS, c. 0%, Sickle Cell is codominant because both traits are expressed when both alleles are present. It appears almost like an incompletely dominant trait because the heterozygous phenotype is intermediate between the fully dominant and fully recessive phenotypes,

	Hb^S	Hb^S
Hb^A	Hb^AHb^S	Hb^AHb^S
Hb^S	Hb^SHb^S	Hb^SHb^S

29. Skin cancer results from damage to the DNA of skin cells, there is no damage to the DNA of the egg cells in the ovary. Her child will not inherit skin cancer.

Sections 19.5, 19.6, 19.7, 19.8

1. karyotype; **2.** autosomes, sex; **3.** Y; **4.** sex-linked; **5.** sex-influenced; **6.** Nondisjunction; **7.** deletion; **8.** translocation; **9.** Human Genome; **10.** males inherit only one X chromosome; they cannot be protected by a normal allele on a second X chromosome; **11.** men receive a Y chromosome from their father; their mother is the only parent that donates an X chromosome; **12.** the father has only one X chromosome—the one with the trait—and he will pass it to all of his daughters; **13.** a. X^hY, b. X^hX^h, c. X^HY, d. X^HX^H or X^HX^h, e. X^HX^h, f. 50%, 50% **14.** The expression of pattern baldness, a sex-influenced trait, is enhanced by testosterone causing it to be expressed more strongly in men than in women; **15.** proteins play many roles in body chemistry that could influence behavior; **16.** Triple X syndrome; **17.** PKU; **18.** cri-du-chat syndrome; **19.** Klinefelter syndrome; **20.** Down syndrome; **21.** Huntington disease; **22.** Jacob syndrome

Looking Back/Looking Forward

1. no; **2.** There is no right or wrong answer to this question; it is your opinion, just be sure you justify your answer

Chapter Test

1. d; **2.** b; **3.** a; **4.** c; **5.** c; **6.** b; **7.** c; **8.** b; **9.** d; **10.** e; **11.** c; **12.** c; **13.** a; **14.** c; **15.** d; **16.** a; **17.** d; **18.** c; **19.** b; **20.** c; **21.** d; **22.** b; **23.** d; **24.** b; **25.** a

Chapter 20

Sections 20.1, 20.2

1. biotechnology; **2.** genetic engineering; **3.** DNA polymerase; **4.** restriction enzymes; **5.** palindromic; **6.** DNA ligase; **7.** plasmids; **8.** polymerase chain reaction; **9.** DNA fingerprinting; **10.** T; **11.** T; **12.** F, single-stranded; **13.** T; **14.** T; **15.** F, polymerase; **16.** F, stops; **17.** T; **18.** T; **19.** T; **20.** T; **21.** T; **22.** F, strand complementary to the original fragment; **23.** billions of years; **24.** they protect bacteria from viral infection by cutting up the viral DNA; **25.** they function as extra-chromosomal DNA in bacteria and contain genes for reproduction, an engineered plasmid can be reinserted in a bacterium and reproduced every time the bacterium reproduces; **26.** copies of DNA produced by PCR lack regulatory genes; **27.** DNA contains repeating nucleotide sequences called junk sequences that are non-coding, the lengths of junk sequences are different in different individuals; **28.** 5; **29.** 3; **30.** 6; **31.** 2; **32.** 1; **33.** 4

Sections 20.3, 20.4

Crossword Puzzle: **1.** retroviruses; **2.** hepatitis B; **3.** fertilized; **4.** engineered; **5.** failures; **6.** ten; **7.** bacteria; **8.** transgenic; **9.** vector; **10.** immuno; **11.** interleukins; **12.** Genome; **13.** liposomes; **14.** difficult; **15.** therapy; **16.** proteins; **17.** nasal; **18.** pharming; **19.** vaccines; **20.** insulin, human growth hormone, erythropoietin; **21.** Viruses made from weakened or killed disease-causing organisms can sometimes cause the disease, while viruses produced in recombinant bacteria contain only the antigen of the disease-causing organism and so carry no risk of causing disease; **22.** inserting recombinant DNA into a sufficient number of cells to correct a defect, and treating a disease without being able to prevent the passing of the disease to offspring; **23.** No, they will still be at risk of passing this disorder to their offspring

Looking Back/Looking Forward

1. yes; **2.** Clean up of toxic wastes and oil pollutants, monitoring hazardous waste sites

Chapter Test

1. c; **2.** c; **3.** d; **4.** d; **5.** a; **6.** d; **7.** b; **8.** a; **9.** c; **10.** b; **11.** b; **12.** b; **13.** d; **14.** a; **15.** d; **16.** d; **17.** d; **18.** d; **19.** a; **20.** d; **21.** d; **22.** c; **23.** b; **24.** d; **25.** b

Chapter 21

Sections 21.1, 21.2

1. acrosome; **2.** head; **3.** midpiece; **4.** tail; **5.** corona radiate; **6.** cytoplasm; **7.** nucleus; **8.** plasma membrane **9.** first polar body; **10.** zona pellucida; **11.** acrosome; **12.** upper third of the oviduct; **13.** vagina, cervix, uterus, oviduct; **14.** 24 hours; **15.** zona pellucida and corona radiate; **16.** entry of the sperm's nucleus into the egg; **17.** sperm entry triggers the release of enzymes that cause changes in the zona pellucida making it impenetrable to additional sperm; **18.** entry of the sperm; **19.** zygote; **20.** when the nuclei of the sperm and mature ovum unite; **21.** follicle cells and polar bodies; **22.** cleavage; **23.** growth;

24. morphogenesis; **25.** fraternal twins; **26.** twins; **27.** differentiation; **28.** conjoined twins; **29.** identical twins

Sections 21.3, 21.4

1. j; **2.** d; **3.** h; **4.** l; **5.** g; **6.** o; **7.** b; **8.** a; **9.** f; **10.** c; **11.** i; **12.** m; **13.** k; **14.** n; **15.** q; **16.** u; **17.** t; **18.** p; **19.** e; **20.** r; **21.** s; **22.** 2 weeks; **23.** morula; **24.** morula; **25.** embryonic disc; **26.** blastocyst; **27.** implantation; **28.** trophoblast cells; **29.** the appearance of an amniotic cavity, and of ectoderm and endoderm in the embryonic disc; **30.** week 3 through week 8; **31.** the appearance of mesoderm; **32.** mesoderm; **33.** endoderm; **34.** ectoderm; **35.** T; **36.** T; **37.** F, amnion; **38.** T; **39.** T; **40.** F, chorion

Sections 21.5, 21.6, 21.7, 21.8, 21.9, 21.10

1. 5th month; **2.** after 8 weeks; **3.** 4th month; **4.** 6th month; **5.** 5th month; **6.** 6 weeks; **7.** 3rd trimester; **8.** 4th month; **9.** 3rd month; **10.** 3; **11.** 1; **12.** 6; **13.** 4; **14.** 5; **15.** 2; **16.** episiotomy; **17.** surfactants, lungs; **18.** ductus venosus; **19.** foramen ovale; **20.** connective; **21.** colostrum; **22.** prolactin, oxytocin; **23.** neonatal; **24.** triple; **25.** immune; **26.** childhood; **27.** GnRH; **28.** childhood; **29.** telomere; **30.** LH, FSH; **31.** a. less elastic, b. stiffen, c. decreases, d. decreases, e. stiffens, f. decreases, g. decreases, h. enlarges; **32.** irreversible; **33.** encodes a protein called testis-determining factor that directs initial development of the testis; **34.** absence of a Y chromosome; **35.** dilation, expulsion, afterbirth; **36.** This embryo will develop as a female. Male development results from testosterone – in the absence of testosterone the embryonic development will default to the female pathway.

Looking Back/Looking Forward

1. Uterine contractions stimulate increased release of oxytocin, which stimulates more uterine contractions, taking the body's response even further away from homeostasis; **2.** increased the growth rate

Chapter Test

1. b; **2.** d; **3.** c; **4.** a; **5.** b; **6.** c; **7.** d; **8.** a; **9.** c; **10.** a; **11.** b; **12.** c; **13.** a; **14.** a; **15.** c; **16.** a; **17.** d; **18.** d; **19.** a; **20.** b; **21.** c; **22.** b; **23.** b; **24.** a; **25.** c

Chapter 22

Sections 22.1, 22.2

1. evolution; **2.** microevolution; **3.** macroevolution; **4.** fossils; **5.** homologous; **6.** analogous; **7.** vestigial; **8.** biogeography; **9.** continental drift; **10.** population; **11.** genetic drift; **12.** bottleneck effect; **13.** founder effect; **14.** gene flow; **15.** extinction; **16.** adaptive radiation; **17.** descent over time, genetic modification, unpredictable, natural; **18.** descent with modification; **19.** one; **20.** sediment, volcanic ash; **21.** stratification; **22.** lowest; **23.** potassium; **24.** vestigial structures; **25.** biochemical molecules; **26.** diverge; **27.** Pangaea; **28.** when two different species evolve in the same or similar environments this can lead to the development of analogous structures because both species will be experiencing similar environmental pressures; **29.** development in the early embryo may be so critical that any changes resulted in death

Sections 22.3, 22.4, 22.5

1. b; **2.** i; **3.** f; **4.** k; **5.** h; **6.** a; **7.** d; **8.** j; **9.** c; **10.** e; **11.** g

Section 22.6

1. Animalia; **2.** eukaryotic, multicellular, complex anatomy; **3.** Chordata; **4.** nerve cord, backbone; **5.** Mammalia; **6.** mammary glands; **7.** Primata; **8.** five digits, flat fingernails; **9.** Hominidae; **10.** upright posture, enlarged brain; **11.** *Homo;* **12.** human form; **13.** *Sapiens;* **14.** complex social structure, complex spoken language; **15.** a; **16.** d; **17.** c; **18.** c; **19.** b; **20.** a; **21.** e; **22.** d; **23.** a; **24.** b; **25.** d; **26.** c; **27.** d; **28.** c; **29.** a; **30.** b; **31.** e; **32.** habitual upright posture and bipedal walking; **33.** hominoids are larger and have bigger brains than primates, hominids are more like humans than chimpanzees

Looking Back/Looking Forward

1. yes; **2.** Adaptive radiation usually follows periods of environmental change that eliminate previous species and create an abundance of new habitats

Chapter Test

1. a; **2.** c; **3.** e; **4.** c; **5.** b; **6.** b; **7.** a; **8.** c; **9.** c; **10.** c; **11.** c; **12.** a; **13.** b; **14.** b; **15.** c; **16.** b; **17.** b; **18.** b; **19.** b; **20.** b; **21.** d **22.** d; **23.** c; **24.** d; **25.** c

Chapter 23

Sections 23.1, 23.2, 23.3

1. ecology; **2.** ecosystem; **3.** population; **4.** community; **5.** biosphere; **6.** habitat; **7.** biotic potential; **8.** environmental resistance; **9.** carrying capacity; **10.** growth rate; **11.** zero population growth; **12.** demography; **13.** habitat **14.** competition for resources, intolerable conditions, physical obstacles; **15.** biotic potential, carrying capacity; **16.** environmental resistance; **17.** increase; **18.** decrease, increase; **19.** decrease, increase; **20.** 2.65; **21.** more; **22.** declining, higher; **23.** equals; **24.** F, populations; **25.** T; **26.** F, each woman; **27.** F, industrial development and economic well-being of the citizens; **28.** F, MIC; **29.** T; **30.** t; **31.** They will be in competition for food, shelter, and resources; **32.** the number of offspring produced by each reproducing member, the length of time required for offspring to reach sexual maturity, the ratio of males to females, the number of reproductive members of the population; **33.** famine, disease, and harsh environmental conditions prevented an increase in the carrying capacity; **34.** the Industrial Revolution; **35.** Not all children will survive to reproductive age; **36.** the LICs; these countries have the fewest resources and the poorest ability to care for a large population

Sections 23.4, 23.5, 23.6

1. climax community; **2.** ecological pyramid; **3.** denitrification; **4.** biomass; **5.** nitrogen fixation; **6.** fossil fuels; **7.** succession; **8.** consumers; **9.** food web; **10.** herbivores; **11.** niche; **12.** decomposers; **13.** omnivores; **14.** biogeochemical cycle; **15.** nitrification; **16.** producers; **17.** carnivores; **18.** photosynthesis; **19.** energy and nutrients; **20.** Many years are required to establish the complex balance between species and resources that exists in a climax community and it cannot be easily restored; **21.** c; **22.** a; **23.** d; **24.** b; **25.** f, j, k, l; **26.** f, j, k; **27.** g, k; **28.** e, h, i; **29.** c; **30.** e; **31.** b; **32.** d; **33.** a; **34.** f;; **35.** destroyed; **36.** chloroplast; **37.** chlorophyll; **38.** NADPH, ATP, oxygen; **39.** Calvin; **40.** energy, carbon dioxide; **41.** carbohydrates; **42.** exchange, biomass; **43.** gaseous; **44.** bacteria; **45.** plants; **46.** nitrogen gas; **47.** sedimentary; **48.** because some usable energy is lost each time energy is transferred from one organism to another; **49.** because the chemicals that compose living organisms are constantly cycled between organisms and the earth without loss of usable matter

Looking Back/Looking Forward

1. This is a question of opinion; **2.** increase

Chapter Test

1. b; **2.** b; **3.** d; **4.** a; **5.** c; **6.** d; **7.** c; **8.** d; **9.** a; **10.** b; **11.** b; **12.** a; **13.** a; **14.** d; **15.** c; **16.** a; **17.** a; **18.** d; **19.** d; **20.** d **21.** c; **22.** c; **23.** d; **24.** b; **25.** d

Chapter 24

Sections 24.1, 24.2, 24.3

1. pollutants; **2.** greenhouse effect; **3.** global warming; **4.** deforestation; **5.** ozone; **6.** desertification; **7.** acid precipitation; **8.** smog; **9.** thermal inversion; **10.** aquifer; **11.** eutrophication; **12.** biological magnification; **13.** burning of vegetation, combustion; **14.** burning of vegetation releases CO_2 from the vegetation and reduces the number of plants available for CO_2 uptake in photosynthesis, combustion releases CO_2 from fossil fuels; **15.** increase; **16.** temperatures may increase; **17.** stratosphere; **18.** troposphere; **19.** decrease; **20.** increase; **21.** sulfur dioxide; **22.** smog formation; **23.** less than; **24.** increase, decreased; **25.** converts freshwater into; **26.** cannot; **27.** more; **28.** do; **29.** global warming, ozone destruction, acid rain, smog production; **30.** chlorine, oxygen; **31.** organic nutrients; **32.** algae; **33.** oxygen; **34.** power plants; **35.** drinking water; **36.** 50; **37.** 50

Sections 24.4, 24.5, 24.6

1. coal, oil, gas; **2.** retrieving and transporting them has an environmental cost; **3.** burning fossil fuels contributes to global warming, acid rain, and smog; **4.** fossil fuels are a nonrenewable resource; **5.** expensive; **6.** generates hazardous waste; **7.** flowing water; **8.** requires a dam which can disrupt the ecosystem; **9.** plant material; **10.** contribute to pollution; **11.** not easily transported; **12.** can only be produced where biofuel crops are grown; **13.** uses land that would otherwise be used for food crops; **14.** Solar power technology is not capable of meeting the energy demand; **15.** number of species, pollution, 1destruction of habitats, exploitation of resources, see sec. 24.5 for additional possible answers; **16.** "development that meets the needs of the present without compromising the ability of future generations to meet their economic needs"; **17.** GDP is the market value of all goods and services produced in a country in one year; while the GPI, in considering the yearly market value, also considers the cost in both environmental and social terms.; **18.** see sec. 24.6 in the text.

Looking Back

1. Energy flows through an ecosystem, with each successive consumer receiving less energy as stated in the second law of thermodynamics. Matter is recycled through an ecosystem. Since toxic pollutants are matter, they will accumulate in tertiary consumers who ingest large numbers of organisms from lower levels of the food chain.

Chapter Test

1. d; **2.** c; **3.** d; **4.** b; **5.** c; **6.** d; **7.** b; **8.** b; **9.** b; **10.** a; **11.** a; **12.** a; **13.** c; **14.** a; **15.** d; **16.** c; **17.** a; **18.** c; **19.** c; **20.** a; **21.** c; **22.** a; **23.** b; **24.** e; **25.** c